ICT and Language Learning:
A European Perspective

T0262922

LANGUAGE LEARNING AND LANGUAGE TECHNOLOGY

Series Editors:

Carol A. Chapelle, *Iowa State University, Ames IA, USA*
Graham Davies, *Thames Valley University, London, UK*

ISSN 1568-248X

ICT AND LANGUAGE LEARNING: A EUROPEAN PERSPECTIVE

EDITED BY:

ANGELA CHAMBERS
AND
GRAHAM DAVIES

SWETS & ZEITLINGER
PUBLISHERS

LISSE   ABINGDON   EXTON (PA)   TOKYO

Library of
Congress Cataloging-in-Publication Data

ICT and language learning : a European perspective / [edited by] Angela Chambers a
Graham Davies.
    p. cm. -- (Language learning and language technology, ISSN 1568-248X 1)
    Includes bibliographical references and index.
    ISBN 9026518099-- ISBN 9026518102 (pbk.)
    1. Language and languages--Study and teaching (Higher)--Technological
innovations—Europe. I. Chambers, Angela. II. Davies, Graham. III. Series.

P53.855 .I25 2001
418'.0071'14--dc21

                                                                    2001020

Cover design: Paula van der Heijdt, Amsterdam, The Netherlands.
Printed by : Krips, Meppel, The Netherlands.

© 2001 Swets & Zeitlinger Publishers b.v., Lisse

ISBN 90 265 1809 9 (Hardback Edition)
ISBN 90 265 1810 2 (Paperback Edition)

Contents

Introduction

Angela Chambers
University of Limerick, IE

While it is accepted that technological developments can lead to profound changes in society and in our daily lives, there is a tendency to underestimate the role of individuals and groups not directly involved in the technological invention itself in determining the ways in which its potential can best be realised. In the area of Information and Communications Technologies (ICT) and language learning in particular, there is a danger that emphasis will be placed on the technological innovation itself, with little or no attention given to the development of theory- or pedagogy-driven approaches to research. Experts and researchers in the field of ICT and language learning are increasingly emphasising that, once a new form of technology has become available, the starting point of research projects should not be the innovation itself but rather its role in the language learning process. As early as 1993 it was noted that there is a need for "sound empirical research into the relationship between second language acquisition and the use of technology on which to base new learning models" (Hagen 1993:108). More recently, in his seminal work which attempts to define CALL as a research area, Levy stresses the need for the CALL community "to build upon what has gone before, rather than be led purely by the latest technological innovation" (Levy 1997:xi). Investigating the interdisciplinary nature of CALL, he underlines the importance of its relationship with a variety of disciplines, including second language acquisition, artificial intelligence, cognitive psychology and computational linguistics. Thus, while a single technological development such as e-mail or Authoring Tools will often form the technological focus of a research project, the success of the project will depend to a great extent on the theoretical or pedagogical approach adopted by the researcher. The chapters in this book are at the forefront of current research in their emphasis on the ways in which the use of technology in language learning may be firmly embedded in a theoretical or pedagogical context.

In an area such as ICT and language learning, which is clearly global in its scope and potential, the emphasis on the European dimension in the title of this work may initially seem surprising. There are, however, a number of reasons why it is particularly appropriate. While language learning is clearly an activity with global relevance, it is seen as particularly important in Europe in both practical and political terms. Multilingualism and language diversity are considered as priorities, not only in the economic sphere as a means of ensuring the mobility of workers, but also in the political context of European integration, as increasing mobility is also seen as a means of encouraging greater interaction and understanding between the different European cultures and states. This makes Europe a particularly fertile ground for collaborative projects in language learning which bring together learners, teachers and researchers from various states, often with financial support from European organisations such as the European Commission. Priority is given to areas which are considered to be particularly effective or to have significant potential, and Information and Communications Technologies figure prominently among these priorities. For example, in *Towards a Europe of Knowledge*, the European Commission gives as one of its aims for developments in the areas of education and training in the twenty-first century the creation of a multilingual "virtual European education area" (1997:5). Collaborative projects in ICT and language learning are thus supported by a number of programmes, the most relevant in the context of the projects included here being the SOCRATES programme.

The majority of the contributors to this volume came together initially as part of one such project, the SOCRATES Thematic Network Project (TNP) in the Area of Languages (1996-1999), co-ordinated by Wolfgang Mackiewicz, Freie Universität Berlin. Most of the chapters form part of the work of the Scientific Committee for New Technologies and Language Learning in the first year of the TNP.[1] In that year a number of key areas were identified where, in the view of the Committee, research projects have an important contribution to make to developments in language learning in higher education in Europe. These include new approaches to language learning and teaching based on learner autonomy, the development and evaluation of Authoring Tools, the contribution of Human Language Technologies to language learning and teaching, and the integration of tools and resources (such as the Web, e-mail, corpora and concordancing, and videoconferencing) into the language learning process. In addition to the work originating directly from the TNP, it should be noted that the chapter by David Little arises from another European project, namely the International E-Mail Tandem Network, and the chapter by David Bickerton, Tony Stenton and Martina Temmerman provides results from the RAPIDO Project, which was part-funded by the SOCRATES (Lingua) programme. Reports on these were included in the work of the TNP in Year 1, as they formed part of the key areas in ICT and language learning identified by the Scientific Committee. These contributions were further complemented by chapters by Jean-Claude Bertin and Mathias Schulze, who joined in the activities of the group at a later stage under the aegis of the European Language Council.

The European context which underpins the chapters in this book may thus be defined as an approach based on collaborative projects involving partners in several countries and focusing on three sets of priorities. Firstly, as language teachers, the partners aim to base their teaching on sound theoretical and pedagogical principles. Secondly, as specialists in ICT and language learning, they aim to integrate technology into their teaching. Finally, in the European context, they aim to make an important contribution to the creation of the virtual European education area referred to above, to complement the work of the physical campuses and to maximise contact between them. Indeed, in the context of developing cross-cultural capability between European states, virtual mobility as an enhancement of physical mobility is playing an increasingly important role (Collis 1997). Needless to say, this European perspective in no way excludes broader international co-operation.

The first chapter by Graham Davies, entitled "New technologies and language learning: A suitable subject for research?", provides a synthesis of recent developments in the definition and recognition of research in this area in higher education, and of the initiatives of international associations such as EUROCALL, CALICO and IALL. The chapter by David Little, "Learner autonomy and the challenge of tandem language learning via the Internet", is a clear example of the integration of research and practice. It is followed by Jeannette Littlemore's study of the link between current theory and practice in higher education in the area of ICT and language learning, which examines to what extent current research is effective in influencing practice in language learning in higher education in Europe. Significantly, both Little and Littlemore present learner autonomy as the theoretical context in which they evaluate the practical application of ICT and language learning. While Littlemore points out the disparity which exists between universities in relation to knowledge about, and attitudes to autonomy, Little's study of the International E-Mail Tandem Network illustrates on the one hand the benefits which accrue to language departments where technical developments are evaluated on the basis of theories of learning, and on the other hand the need for research and evaluative studies to accompany practical developments.

The three chapters on authoring which follow may appear at first sight to represent a shift away from this theory-based approach, but a closer reading reveals that they are also motivated by the need to integrate Authoring Tools into a theory- and pedagogy-driven environment. The study of "Criteria for the evaluation of Authoring Tools in language education", by David Bickerton, Tony Stenton and Martina Temmerman, begins by noting the plethora of conflicting claims made in relation to how easy or difficult it should be to use an Authoring Tool. They then propose an evaluation system which has been developed in the context of the RAPIDO Project, and which will be of considerable interest and use both as a guide for users of Authoring Tools who are not active researchers in the area, and also for researchers, as an important initiative in the development of further evaluation systems for Authoring Tools. It is significant that in their conclusion the authors present this research as a necessary first step in the development of more comprehensive evaluation systems which focus on pedagogical issues. Philippe Delcloque's study, "DISSEMINATE or not? Should we pursue a new direction: Looking for the 'third way' in CALL development?", proposes "authorability" as a key element in the development of both authoring tools and courseware, and presents the concept and architecture of DISSEMINATE as a flexible and unifying element in the current complex scenario in CALL. He is also motivated by the need to move away from a technology-driven environment to one where linguistic and pedagogic issues will transcend technological constraints. Jean-Claude Bertin's chapter on "CALL material structure and learner competence", based on the system which the author himself developed, provides an excellent example of the advantages of taking as a starting point not the technological innovation, but the theoretical area relevant to its use, in this case the learners' cognitive profiles. In this chapter the issues involved in the choice between freedom and guidance in the design of materials are explored in depth from both a theoretical and practical perspective.

The two chapters on what is known as Language Engineering (LE), Natural Language Processing (NLP) and, more recently, Human Language Technologies (HLT) play an important role in the volume, particularly for readers who have considerable knowledge and experience of CALL, but not of HLT, and who are aware of the increasing importance of this area within CALL. In his chapter entitled "From gap-filling to filling the gap: A re-assessment of Natural Language Processing in CALL", Sake Jager begins by admitting that NLP has lived a somewhat precarious existence to date in the context of CALL, and then argues that a focus on the learners' needs, on methodology and on the role of the teacher is necessary in NLP research if it is to reach its full potential in the context of CALL. Mathias Schulze's study of "Human Language Technologies in Computer-Assisted Language Learning" gives a broad overview of the applications of HLT in CALL, and argues that advances in HLT have already made an important contribution to CALL, in particular to the learner's control over the communicative interaction. He concludes that developments in CALL in the future will continue to benefit greatly from research in HLT.

The chapters which follow deal with aspects of ICT and language learning as diverse as the Internet, concordancing and videoconferencing. Thomas Vogel, in his chapter on "Learning out of control: Some thoughts on the World Wide Web in learning and teaching foreign languages", emphasises that the sometimes extravagant claims made for the Web in the context of language learning need to be supported or disproved by empirical research. He investigates the extent to which the Web can be said to constitute a naturalistic language learning environment, and concludes that, at this early stage in its development, the Web has raised as many problems and questions as it has provided answers. In his study of concordancing, entitled "Concordances in the classroom: The evidence of the data", Joseph Rézeau investigates the potential of the use of corpora and concordancing software in providing language learning and teaching resources based on authentic data. Another researcher in this area has described the concordancer as technologically rudimentary, but very powerful as a cognitive tool (Wolff 1997:22-23), thus underlining once again the potential for research such as that described in this chapter. Rézeau's study reveals how a tool as simple as the concordancer can be used, particularly at a time when access to corpora is increasing dramatically, as a means of access to real language use instead of the often artificial environment of the invented examples in the manual or the simplification necessary in the limited format of the grammar or the dictionary. Finally, in "ReLaTe: A case study in videoconferencing for language teaching", John Buckett and Gary Stringer provide another example of the research and development work necessary to underpin the application of what might at first sight appear to be a relatively straightforward technological development, in order to meet the needs of language learners and teachers. They stress the importance of such developments for the provision of specialist courses and minority languages in particular.[2]

The eleven chapters thus provide, in their different ways, responses to the various documents referred to in the first chapter which stress the need for greater integration of research and practice in CALL. They also reveal that, despite the increasing number of publications in ICT and language learning, research in this area can still be described as being at a very early stage of development, if not in its infancy. It is not surprising that many of the chapters conclude with a look to the future, stressing the need to build on what has already been achieved. Among the areas identified as requiring further development, the most prominent are the need for controlled empirical studies and the need for greater ease, accessibility and flexibility in the technology which is available, so that the users, be they learners, teachers or researchers, can concentrate on the linguistic, pedagogic and educational challenges which face them.

Notes

[1] For a synthesis of the work of the Committee from 1996-1999, see Chambers (1999).

[2] A longer version of this study was produced as part of the activities of the first year of the TNP and is available (2000) at:
http://www.odur.let.rug.nl/projects/tnp-ntll/TNP_report.doc

References

Towards a Europe of Knowledge, 2000-2006 (1997). Communication from the Commission COM (97) 563. Luxembourg: Office for the Official Publications of the European Communities. Also available (2000) at: http://europa.eu.int/en/comm/dg22/orient/orie-en.html

Chambers, A. (Ed.) (1999). *Information and Communications Technologies and language learning: Linking policy, research and practice*. Available (2000) at: http://www.userpage.fu-berlin.de/~elc/TNPproducts/SP3report.doc

Collis, B. (1997). New technologies and internationalism: an overview of models. Paper presented at the Nuffic seminar on Virtual Mobility: New Technologies and Internationalisation, May 1997, University of Twente, the Netherlands. Available (2000) at: http://www.nuffic.nl/gateway/vm_seminar/collis.html

Hagen, S. (1993). Technology and learning. In *Foreign language learning and the use of new technologies* (pp. 108-112). Brussels: Bureau Lingua.

Levy, M. (1997). *CALL: Context and conceptualisation*. Oxford: Oxford University Press.

Wolff, D. (1997). Computers as cognitive tools in the language classroom. In A-K. Korsvold & Rüschoff, B. (Eds), *New technologies in language learning and teaching* (pp.17-26). Strasbourg: Council of Europe Publishing.

1

New technologies and language learning: A suitable subject for research?

Graham Davies
Thames Valley University, UK

1. Definition of terms

The nature and recognition of research in the area of technology and language learning raises a number of fundamental questions. First, the term "new technologies" needs to be defined. What is new today is old hat six months later. It is generally agreed, however, that the term "new technologies" in the context of language learning and teaching – for the time being at least – is concerned with technologies in which the computer plays a central role, e.g. Computer-Assisted Language Learning (CALL) and the use of the Internet in promoting language learning and teaching. For the sake of convenience and brevity, the term "CALL" is used in this chapter, as defined in the *EUROCALL, CALICO, IALL Joint Policy Statement* (1999): an "academic field that explores the role of information and communication technologies in language learning and teaching" (see Appendix).

2. Finding a research topic in "the good old days"

In "the good old days" it seems to have been relatively easy to find a research topic and to get it recognised. One member of staff in the German department of the university where I studied both as an undergraduate and as a postgraduate in the 1960s gained his PhD by compiling a concordance to the works of Stefan George. He would not be able to get away with that today as a computer could probably do the job in a matter of hours – or even minutes. My own research involved the compilation of a lexicon of Medieval German heraldic and tournament terminology based on original sources. Ironically, this is what led me into my first contact with computers. As I set about hacking the mass of notes and index cards that I had accumulated into an intelligible format, I found myself faced with an organisational problem that I was eventually able to solve by using computer technology. After that I never, literally, looked back.

I clearly recall the process of selecting my research topic. It began with a friendly chat with my supervisor who threw out a few suggestions. I did some background reading, found an area that seemed to have been untouched by other researchers and narrowed it down, agreeing the final topic with my supervisor and then getting down to work. The research itself involved many, many days spent in the British Museum Library poring over every Medieval German text that I could lay my hands on. This was research in the traditional sense.

3. Finding a research topic today

The university at which I currently supervise two postgraduate students has a much more rigid procedure in place. Every research degree topic has to be approved by the Research Degrees Committee (RDC). This means that prospective students have to do a substantial amount of preliminary research before they can be officially enrolled as postgraduate students. The friendly chat with the supervisor and the selection of a topic in agreement with the supervisor are part of the process, but then the serious work starts and the prospective student has to write a lengthy proposal which is scrutinised by the RDC. The RDC consists of senior academics from the main subject areas in the university, and the student's proposal has to be intelligible both to specialists and non-specialists – which is not always an easy task. I recall a lecturer in one of the university's science departments being puzzled by a proposal submitted by one of my students, Yu Hong Wei. Part of her research involved examining a commercially available grammar checker program in order to establish if it was effective in identifying errors in authentic essays written by students of English as a Foreign Language (EFL). "Surely," the science lecturer commented, "you can't base research on checking that a computer program works. All computer programs have to work, don't they?" I had to point out that in the area of languages this is not necessarily so. The topic was, however, approved by the RDC, and useful interim results have been published (Yu Hong Wei and Davies 1997).

I have been personally involved in a wide range of research topics in the area of new technologies and language learning, some of which have already led to the award of PhD degrees and some of which are still in progress. Topics include:

- The creation of a corpus aiming to assist students to develop technical writing skills.
- An inquiry into the conceptual frameworks and working methods of CALL authors.
- The design of a set of computer programs based on schemata theory to assist students to develop effective reading skills.
- A study on the effectiveness of pre- and post-writing tools.
- The design of an efficient Cloze package to encourage behaviour commensu-

rate with the role of Cloze in a theory of effective reading strategy.
- A study on the effectiveness of computer programs in assisting with vocabulary acquisition.

4. Recognition of research in the area of new technologies

The key problem associated with research in the area of new technologies is one of recognition. A good deal of research in CALL is closely associated with development projects and therefore goes unrecognised as "real" research. Dodigovic (1998) summarises the problem as follows:

> The reasons that research is not uniformly recognised as a constituent part of CALL software development projects can be sought both within and without the CALL profession. Outside the profession, in the sciences, research and development tend to be viewed as separate entities, development following upon research as a simple application process (Valter 1988). For example, the *Macquarie Dictionary* (1996) defines research and development as "that part of industry concerned with scientific research and the technological development of the results." Perhaps it is this very distinction that polarises CALL developers themselves. In his extremely thoughtful essay on theory driven CALL, Levy (1997b) finds that CALL development is frequently divided between those who make a particular theory their point of departure (formalists) and those who make discoveries while writing programs (proceduralists). The former appear to believe that completed research is a prerequisite for successful development, whereas the latter allow for the integration of research into the development process. Levy acknowledges that revision of original theory is a recurrent factor when development is undertaken in a multidisciplinary area such as CALL. Knowledge gained in the course of development may confirm or disconfirm various aspects of an underlying theory, thereby altering and improving the original theory itself. Clearly, the research process is far from complete when a theory is selected to serve as a basis for a CALL development project. (Dodigovic 1998:26)

Maintaining a clear distinction between *research* and *development* CALL is not always easy, and is often deliberately obscured by practitioners due to the constraints of their funding organisations, some of which favour "pure" research and some of which favour "pure" development. The third element that comes into play is *practice*. The interrelationship between these three elements and the need to make a distinction between them is highlighted in the *EUROCALL, CALICO, IALL Joint Policy Statement* (1999): see Appendix.

So what kinds of research in CALL are desirable and acceptable? It is only since the early 1980s that research in this area could be carried out on a reasonable scale. Prior to this, computers were accessible only to a selected few students in universities, and it was difficult to find a sufficient number of language learners to enable a research project to be set up. When microcomputers began to appear in educational institutions in the early 1980s, they were greeted with a good deal of criticism by conservative members of the teaching profession. Inevitably – and justifiably – questions were asked: How effective is CALL in promoting language learning? What does CALL offer that is different from other modes of learning? Can you prove that computers have something worthwhile to offer? It was up to the researchers to provide some of the answers. A good example of empirical research is described by Trippen, Legenhausen & Wolff (1988). Focusing on strategies adopted by a group of German MT pupils, who were observed while completing total-cloze text-reconstruction exercises in English – an activity which was developed specifically for the computer – the authors were able to establish a hierarchy of favoured strategies:

- *Frequenzstrategien*: looking for high-frequency words: "the", "a", "is", "in", etc.
- *Formale Strategien*: using punctuation clues and counting the blanks representing missing letters.
- *Semantische Strategien*: looking for semantically linked words.
- *Gedächtnisstrategien*: remembering the original text – if it had been looked at before.
- *Grammatische Strategien*: applying knowledge of grammatical rules, e.g. single subject = single verb form ending in "s".
- *Weltwissenstrategien*: general knowledge about the world, e.g. a text about superstition might contain something concerning black cats or ladders.
- *Textuelle Strategien*: looking for words appropriate to the type of text: formal text, dialogue, etc.
- *Ratestrategien*: guessing.

(Trippen, Legenhausen & Wolff 1988:85)

Significantly, guessing was the least-used strategy, thereby confounding teachers who criticised such exercises by dismissing them as guessing games involving very little thinking on the part of the learner.

Recent examples of empirical research are described by Grace (1998), whose work focused on differences in the ability of different personality types in retaining vocabulary while using computer programs, and Myles (1998), who aimed to establish what design features needed to be incorporated into vocabulary acquisition programs in order to maximise their effectiveness. Sadly, Susan Myles passed away before her research was completed.

Second Language Acquisition (SLA) is widely recognised as one of the "respectable" areas of CALL research. Garrett coined the acronym CARLA – Computer-Assisted Research on Language Acquisition – and argued the case for expanding research in CALL beyond evaluating the efficacy of technology:

... we need to understand the differences between all those kinds of research and technology-based inquiries into the psycholinguistic processing that takes place in language learning – and most of all we need to develop the theoretical structure supporting the latter so as to see why and how technology can provide us with insights into that processing that we cannot achieve with any other research methodology. Garrett (1991:19)

Computers, Garrett points out (1991:18), provide us with the means of "tracking learner processes" and "with the means to engage students in radically different language learning behaviours.

Similar points were made again and developed further by Garrett (1998) in the keynote paper she presented at the EUROCALL 97 conference in Dublin, the theme of which was "Where research and practice meet".

Since the 1980s CALL has moved on in leaps and bounds and, as Levy (1997a) points out, citing Chapelle (1989), it is no longer adequate to ask naïve questions concerning the effectiveness of CALL:

Firstly, it is now recognised that the term CALL covers a range of activities, not just one type; next, "second language competence" is now defined as a complex set of interrelated competencies, making it more difficult to test directly as a result; thirdly, researchers have recognised the importance of studying the processes of learning, causing research that focuses on learning outcomes alone to be inadequate; and finally, individual student characteristics have been shown to have a significant impact on SLA. (Chapelle 1989:7–9, cited by Levy (1997a:30))

The range of activities in CALL is significantly wider than it was in the 1980s, and it continues to expand at a breathtaking pace. For example, learner autonomy, self-access and distance learning, all of which are closely bound up with new technologies, have become a focus of attention in recent years (see Little and Littlemore in this publication). The need for more research in these areas was discussed by the New Technologies group at the launching conference of the European Language Council in Lille (Chambers 1999). Wolff (1997:73–75) referred to a number of other research issues that need to be addressed in his EUROCALL 96 paper:

- Motivation: "There has been not one longitudinal study about the motivational potential of computers in the language classroom, we don't know how long motivation lasts."
- The Internet: "Is it true that by reading superficially through texts or by looking at film clips or listening to voices on the World Wide Web items of a foreign language are learnt? Is this type of learning – some call it incidental learning – suited for language learning?"
- Multimedia: "Can we be sure that by bombarding our students with colours,

sounds, and movements on the screen they will learn better than when they are presented with an item in one perceptual mode only?"
- Screen design: "We have to do more research with respect to the screen design we are using."

It is clear that there is ample scope for research in CALL. As the technology develops, the potential areas for research increase.

5. Publications

It used to be difficult to get articles about CALL published, the main reason being that it was not clear where they belonged. The editorial boards of journals special-ising in modern language teaching tended to be sceptical about new technologies, while the boards of journals that focused mainly on the technologies themselves had little understanding of language teaching methodology. On the whole, how-ever, journals with a technology focus tended to be more open-minded. There is now a range of journals that bridge new technologies and language learning and teaching, for example:

- *CALL* (Swets and Zeitlinger, The Netherlands)
- *CALICO Journal* (CALICO, USA)
- *IALL Journal* (IALL, USA)
- *ReCALL* (CTICML/EUROCALL – now taken over by Cambridge University Press)

A recent phenomenon has been the appearance of journals published exclusively on the World Wide Web:

- *Language Learning and Technology Journal*: http://polyglot.cal.msu.edu/llt
- *ALSIC*: http://alsic.univ-fcomte.fr
- *ON-CALL* and *CALL-EJ On-line* (recently merged): http://www.cltr.uq.edu.au/oncall & http://www.lerc.ritsumei.ac.jp/callej/index.html

Web journals are often regarded with suspicion by academics, mainly because the Web is perceived as ephemeral and less permanent than the printed word. Web journals, however, often operate a stringent refereeing process and are no less re-spectable than their printed counterparts.

Books on CALL have been produced steadily since the early 1980s. A compre-hensive CALL bibliography is available at the CTICML website, University of Hull: http://www.hull.ac.uk/cti/resources/reading/callbib.htm

6. The status of CALL research

In spite of the amount of work being done worldwide in the area of new technologies and language learning and teaching, the question of research remains a contentious issue. It is clear that the situation is not all that different in most countries in which new technologies have become widely available in recent years. Lecturers who carry out CALL research or development work often find themselves sitting uncomfortably in a twilight zone which has yet to gain official recognition. In the UK the body that conducts a regular survey of research – the Research Assessment Exercise (RAE) – in the higher education sector and awards funding on the basis of its findings defines "research" as follows:

> "Research" for the purpose of the RAE is to be understood as original investigation undertaken in order to gain knowledge and understanding. It includes work of direct relevance to the needs of commerce and industry, as well as to the public and voluntary sectors; scholarship*; the invention and generation of ideas, images, performances and artefacts including design, where these lead to new or substantially improved insights; and the use of existing knowledge in experimental development to produce new or substantially improved materials, devices, products and processes, including design and construction. It excludes routine testing and analysis of materials, components and processes, e.g. for the maintenance of national standards, as distinct from the development of new analytical techniques. It also excludes the development of teaching materials that do not embody original research.
> *Scholarship for the RAE is defined as the creation, development and maintenance of the intellectual infrastructure of subjects and disciplines, in forms such as dictionaries, scholarly editions, catalogues and contributions to major research databases.
> (RAE Circular 4/99, Appendices, Annex A:
> http://www.rae.ac.uk/Pubs/4_99/4_99.htm)

Jamieson (1998) describes the situation with regard to new technologies and modern languages in her report on the CILT Research Forum, Homerton College, Cambridge, 6–7 January 1998. The relevant section of her report is worth quoting in its entirety:

> The concern about the validity of IT projects as compared to the more traditional research was highlighted in another plenary session, 'The Research Assessment Exercise and its message in relation to IT and foreign languages' delivered by Professor Richard Towell from the University of Salford, a member of a language RAE panel and the chair designate of UCML. The speech outlined the criteria for accepted research as:
> · Research builds on existing knowledge.

· Research takes place within a definable theoretical research framework.
· Research follows a recognisable methodology which allows meaningful statements to be made.
· Good research moves the field forward by providing replicable, verifiable, generalisable results.

New teaching materials and approaches per se are not readily accepted as research; instead projects need to:

· be embedded in, for example, the theory of second language acquisition;
· clearly state the method of investigation;
· analyse the results in such a way that general principles can be deduced.

Some recognition by the new researchers of the validity of these criteria and by RAE panels of the value of this form of research is therefore essential for the next round of the RAE exercise. (Jamieson 1998:133–134)

This was not the first time that this issue had been raised in public. The EUROCALL 97 conference, which took place at Dublin City University, included a seminar on research led by David Little. The seminar was well attended, indicating the level of concern among EUROCALL members. The key issues are summarised by Little (1998) as follows:

First, there had been discussion of the general orientations appropriate to research in CALL. One group suggested that we need to draw on the theories and research practice of other disciplines, including linguistics, psychology, social sciences, anthropology and education; while two groups noted that it is important to be clear what kind of research we intend to engage in and to adopt an appropriate methodology. Research takes time, which costs money, and one group pointed out that without research funding it is impossible to undertake large-scale empirical projects. Secondly, most groups spent some time discussing the implications of a fact to which Nina Garrett drew attention in her opening plenary: that most CALL applications allow researchers to gather large quantities of data with minimum effort. One report pointed out that it is one thing to collect data and another to know what to do with it, and several groups emphasised the importance of good research design. Thirdly, the groups addressed the role that EUROCALL might play in helping to develop a research culture appropriate to CALL. It was suggested that EUROCALL should establish a register of research activities and perhaps a special interest group for research; join forces with CALICO to found a world-wide electronic journal for CALL research; seek funding to sponsor research projects run by its members; organise a summer school on research in CALL; establish an electronic discussion forum on research in CALL; and lobby against the exclusion of research from EU-funded programmes such as SOCRATES and LINGUA. (Little 1998:128)

Following this seminar, the EUROCALL Executive Committee drew up a plan of action in order to help improve the status of CALL research. As a first step, EUROCALL joined forces with IALL and CALICO, two of the leading associations in the USA, and organised a colloquium on research at the University of Essen, Germany, April/May 1999. The colloquium was attended by 20 leading CALL theorists, researchers, developers and practitioners from Europe and the USA. Three important outcomes of the colloquium were:

- the drafting of a policy document which aims to define what CALL research is all about and how it should be evaluated (see Appendix)
- the compilation of a list of higher education institutions that offer postgraduate diplomas or degrees that bridge languages and information and communications technologies:
  http://ourworld.compuserve.com/homepages/GrahamDavies1/courses.htm
- the compilation of a register of research projects at the EUROCALL Deutschland website: http://www.uni-essen.de/~lan501/Eurocall.htm

The Essen colloquium was an important landmark. Whatever other academics and funding bodies may think, there is a considerable consensus of opinion among professionals engaged in research or development in the broad area of new technologies and language learning and teaching. At least the experts agree, and the situation seems to be improving as far as the status of CALL is concerned. A number of professorships have already been conferred upon academics who work predominantly in this area.

7. The politics of CALL

It is significant how CALL has become a political issue in recent years – political in the context of individual educational institutions' internal politics, political in a national context, and political in the wider context of the European Union. Politics in all these contexts is – as usual – bound up with money. The need for CALL practitioners to be aware of the political context in which they operate was addressed by Chesters (1999) in his EUROCALL 98 keynote paper.

Educational institutions' internal politics and the indifferent attitude to new technologies and language learning displayed by national funding bodies concerned with the non-scientific disciplines have already affected the behaviour of CALL practitioners. Some have chosen to "cross to the other" side; they have moved from language departments over to computer science or educational technology departments, where their work is given more credit and is more likely to attract research and/or development funding. This represents a significant loss to language departments. There are even confirmed reports of lecturers in language departments being told in no uncertain terms by their superiors to stop "playing with computers" and to get back to "serious research" into linguistics or literature in order to keep the department's research profile high and to bring in commensurate funding. This attitude is damaging to CALL's reputation, but as Chesters (1999:12) points out: "Ignore dinosaurs [...] in time, they became extinct".

Money is unquestionably the key issue at European level. As Little (1998:128) points out, funding for research is not available to applicants seeking support for development projects in the area of new technologies and language learning and teaching under the European Union's SOCRATES and LEONARDO programmes, which are now managed by the Education and Culture Directorate-General (formerly known as DGXXII). It is assumed that all the research has already taken place by the time the application is made, which – as anyone knows who has taken part in such a project – does not make sense, as important discoveries often go hand in hand with development. There is a remarkable difference in the attitudes adopted by different EC departments, particularly the Education and Culture Directorate-General (the old DGXXII) and the Information Society Directorate-General (the old DGXIII). The former tends to have a negative attitude towards research, while the latter is positive. The Information Society D-G is much more likely to fund a project that includes a substantial research element than the Education and Culture D-G. An example of a funding programme that embraces a wide range of ICT/language activities is referred to by Chesters (1999:9): MLIS (Multilingual Information Society). The MLIS programme was set up by the former DGXIII and covers a wide range of ICT/language R&D areas, from computer-assisted learning to speech recognition, including many areas which fall within the definition Language Engineering – now officially renamed Human Language Technologies: see Schulze's paper in this publication, the HLT website at http://www.linglink.lu, and the following two reports:

i.   European Commission, DGXIII (1998). *Language Engineering: Progress and prospects '98*, Telematics Applications Programme. Luxembourg: LINGLINK, Anite Systems.

ii. European Commission DGXIII (1998). *The EUROMAP Report: Challenge and opportunity for Europe's information society*. Luxembourg: LINGLINK, Anite Systems.

Language engineers, as Chesters (1999:10) indicates, are likely to be "an extremely important, even vital, set of colleagues" for CALL practitioners, especially when they are submitting funding applications: see Schulze, Hamel & Thompson (1999) and the papers by Schulze and Jager in this publication. How, then, should the concerned CALL practitioner behave in the face of these political pressures? Chesters (1999:12) has some important advice, which is worth repeating here:

• Be aware
• Ignore dinosaurs
• Champion champions
• Infiltrate

- Organise
- Seek alliances

3. Conclusion

The answer to the question raised in the title of this chapter is undoubtedly "Yes, CALL is a suitable subject for research". There is ample evidence of ongoing research at the highest academic level, professional associations have been set up, specialised journals are well established, and there is even a worldwide movement underway: WorldCALL. WorldCALL held its inaugural conference at the University of Melbourne in July 1998, and the selected papers have been published (Debski & Levy 1999). All in all, the future of CALL and CALL research looks rosy.

References

Chambers, A. (Ed.) (1999). *Information and Communications Technologies and language learning: Linking policy, research and practice.* Available (2000) at: http://www.userpage.fu-berlin.de/~elc/TNPproducts/SP3report.doc

Chapelle, C. (1989). CALL research in the 1980s: Setting the stage for the 1990s. *CALL Digest, 5.7,* 7–9.

Chesters, G. (1999). On the politics of CALL. *ReCALL, 11.1,* 7–12.

Debski, R. & Levy, M. (Eds) (1999). *WorldCALL: Global perspectives on computer-assisted language learning.* Lisse: Swets and Zeitlinger.

Dodigovic, M. (1998). Elements of research in CALL development projects. *CALICO Journal, 15.4,* 25–38.

European Commission, DGXIII (1998). *Language Engineering: Progress and prospects '98,* Telematics Applications Programme. Luxembourg: LINGLINK, Anite Systems.

European Commission DGXIII (1998). *The EUROMAP Report: Challenge and opportunity for Europe's information society.* Luxembourg: LINGLINK, Anite Systems.

Garrett, N. (1991). Where do we go from here – and who is leading the way? In H. Savolainen & Telenius, J. (Eds), *EUROCALL 91: Proceedings* (pp. 17–20). Helsinki: Helsinki School of Economics.

Garrett, N. (1998). Where *do* research and practice meet? Developing a discipline. *ReCALL, 10.1,* 7–12.

Grace, C. (1998). Personality type, tolerance of ambiguity, and vocabulary retention in CALL. *CALICO Journal, 15.1-3*, 19–45.

Jamieson, A. (1998). Report on the CILT Research Forum, Homerton College, Cambridge, 6-7 January 1998. *ReCALL, 10.1*, 133–135.

Levy, M. (1997a). *CALL: Context and conceptualisation*. Oxford: Oxford University Press.

Levy, M. (1997b). Theory-driven CALL and the development process. *CALL, 10.1*, 41–56.

Little, D. (1998). Report on Seminar on Research in CALL, EUROCALL 97. *ReCALL, 10.1*, 127–128.

Myles, S. (1998). The language learner and the software designer: A marriage of true minds or ne'er the twain shall meet? *ReCALL, 10.1*, 38–45.

RAE Circular 4/99, Appendices, Annex A. Available (2000) at: http://www.rae.ac.uk/Pubs/4_99/4_99.htm

Schulze, M., Hamel, M-J. & Thompson, J. (Eds) (1999). *Language processing in CALL. ReCALL* Special Edition. Hull: CTI Centre for Modern Languages, University of Hull.

Trippen, G., Legenhausen, L. & Wolff, D. (1988). Lernerstrategien und Lernprozesse bei der Bearbeitung von CALL-Software. In W. Kühlwein & Spillner, B. (Eds), *Sprache und Individuum*, (pp. 83–86) Tübingen: Gunter Narr Verlag.

Valter, Z. (1988). Main features of research work in Croatia. In K. Potthast (Ed.), *Proceedings of 4th Scientific Colloquium: Science for Practice*, (pp. 65–80) Bremen: HSB Bremen.

Wolff, D. (1997). Computers and new technologies: Will they change language learning and teaching? In J. Kohn, Rüschoff, B. & Wolff, D. (Eds), *New horizons in CALL: Proceedings of EUROCALL 96* (pp. 65–82). Szombathely, Hungary: Dániel Berzsenyi College.

Yu Hong Wei & Davies, G.D. (1997). Do grammar checkers work? In J. Kohn, Rüschoff, B. & Wolff, D. (Eds), *New horizons in CALL: Proceedings of EUROCALL 96* (pp. 169–188). Szombathely, Hungary: Dániel Berzsenyi College.

Appendix

EUROCALL, CALICO, IALL
Joint Policy Statement arising from a
Research Seminar on CALL
30 April to 1 May 1999, University of Essen, Germany

Introduction

Computer-Assisted Language Learning (CALL) is a relatively new and rapidly evolving academic field that explores the role of information and communication technologies in language learning and teaching. It includes a wide range of activities and initiatives in materials development, pedagogical practice, and *research*. CALL as a field began when the limitations of the computer narrowly limited the pedagogy that could be implemented with it, and consequently some people still believe that CALL refers only to drills and mechanical exercises. Today, however, CALL includes highly interactive and communicative support for listening, speaking, reading and writing, including extensive use of the Internet. Materials development, pedagogy and research have developed in intellectual sophistication to the point where the status of CALL as an academic field of studies requiring special consideration should be seriously considered. CALL is no longer either a straightforward pedagogical application of a new medium, nor simply a practical extrapolation of theoretical work in some other discipline.

CALL is sometimes regarded simply as a sub-section of Computer-Assisted Learning (CAL), but because CALL deals specifically with language learning it is both inherently multidisciplinary and academically substantive. It can be said to belong to the field of Applied Language Studies and, within that, is most closely related to Second Language Acquisition (SLA), which is itself a rapidly evolving discipline. CALL and SLA are related to sociolinguistics, pragmatics, discourse analysis, and psycholinguistics. In addition, CALL is related to work in education, computer science, natural language processing, cognitive science and psychology, linguistics, cultural studies, and media/communication studies. It is influenced by, and in turn influences, theory and research in all these related fields.

The aim of this document

This document has been drafted by a group of twenty CALL theorists, researchers, developers and practitioners from Europe and the USA in order: (i) to establish a clearer understanding for departments, institutions, professional associations and decision-making bodies of the range of activities represented in the field, and (ii) to provide an organised and consistent perspective on the rubrics under which these activities should be evaluated. Assessment and academic recognition of work in

CALL presents difficulties not only because of the rapid evolution of the field but also because of the emergence of new theoretical and methodological paradigms.

CALL work can be categorised as *research, development,* and *practice.* Research may be separate from development, in that a researcher may explore the effects of using technology-based tools or materials developed by others, e.g. formative evaluations, or may focus entirely on theory development. In CALL the progression often begins with pedagogical practice or learner needs driving the development of technology-based materials, techniques, or environments. This development effort may then later lead to research, which in turn may or may not be used to generate theoretical implications. Nonetheless, in establishing criteria for evaluating CALL work for purposes of academic recognition and reward, it is important that the distinctions between these activities be clearly articulated.

Academic standards in CALL

In CALL the term *development* may refer to the creation of pedagogical materials (including design, programming, and incorporation of actual lesson content), or the development of tools and applications into which others can insert language content. In academic contexts where the development of pedagogical materials is typically not rewarded, CALL development is often portrayed as *research* especially when it is based on previous research and/or includes formative research, in which the materials are tried out on learners and feedback is sought as part of the developmental process. Conversely, however, some research projects exploring the feasibility or validity of technology use in language learning are labelled as *development* when funding agencies explicitly proscribe research because they want to support the creation of immediately applicable pedagogical materials. Appropriate evaluation of CALL development work depends crucially on the recognition that not only technical expertise and pedagogical expertise are required – both of a high order – but that in addition this work is a kind of professional activity that is without precedent in the field of language education, and not simply more time-consuming than creating exercises or reading materials. Evaluation of such work must be done by those who can distinguish the levels of expertise required.

When *research* is mentioned in connection with CALL, the assumption is usually that the term refers to studies of the efficacy of technology use in a language learning task that would otherwise be undertaken without it. Data collection and analysis in CALL research may be qualitative or quantitative, experimental or ethnographic, and is published in CALL journals and those of related fields, which naturally include very respected Web-based journals. Crucially, of course, CALL research also includes developmental and prototypal computing. CALL research is moving into new areas, drawing on theories from related fields and creating its own theoretical and methodological paradigms. It is indeed a sign of maturity that CALL has now standardised its terminology, identified its points of reference, and includes a significant number of sub-branches of activity. The design expertise required is of an entirely different kind than that involved in the development of conventional pedagogical materials.

An example of CALL research that is recognised as academically valid is the use of data collected while students are using technology-based materials to confirm or disprove hypotheses generated by SLA theory, whether sociolinguistic or psycholinguistic. This kind of CALL research can contribute to the development of CALL theory itself, i.e. to the understanding of how technology use actually changes the process of language learning, and is thus a crucial part of the paradigm shift needed to establish CALL as a discipline in its own right. In fact, the *process* orientation of much current SLA theory and research depends crucially on CALL research.

The evaluation of pedagogical practice, materials development, and research in CALL can be based on assessment mechanisms as objective as those used in other fields, but it requires an understanding of the particular challenges of CALL that is not yet widespread in language departments and academic institutions. Interdisciplinarity and paradigm shifts always make evaluation problematic. This document attempts to set out some of the crucial considerations.

Draft 6.0, 13 October 1999

This document will be regularly revised in the light of new developments. The most recent version is available at the EUROCALL website:
http://www.eurocall.org

2

Learner autonomy and the challenge of tandem language learning via the Internet

David Little
Trinity College Dublin, IE

1. Introduction: Self-access and learner autonomy

Over the past two decades the introduction of self-access systems has been the single most important development affecting the learning of foreign languages in universities around the world. This development took its first impulse from the recognition that language laboratories could as well be used by individual students working on their own as by classes of students closely monitored by a teacher. In time the framework for open learning thus created came to accommodate other technological aids, such as video playback facilities, satellite television, and computers. Originally self-access systems were set up to support language learning; more recently the development of the Internet as a channel of communication has given them the capacity to support language learning through language use.

The fact that in self-access systems language learners mostly work on their own has led many hard-pressed universities to see them as a way of saving money: buy machines and learning materials now and they will serve you for five years; hire teachers now and you must pay them for as long as you offer the courses they staff. Even in terms of simple economics, of course, this view is hopelessly naïve; for if a self-access system is to succeed, it requires a great deal more than initial capital investment. The system itself must be designed according to the particular needs of the larger environment it is to serve; it must then be manned; its technical installations must be maintained; its learning materials must be regularly updated; and provision must be made to advise users how to go about their language learning. Each of these things costs money.

The view of self-access systems as a means of saving money goes hand in hand with a tendency to treat "self-access" and "autonomy" as interchangeable concepts: self-access systems are assumed to be places where learners work without a teacher, and when students work without a teacher they are assumed to be autonomous (Little 1997). In the theory of adult education, however, autonomy has long been defined not organisationally, but in terms of the learner's psychological relation to the content and process of learning – a relation that is founded on a capacity for detachment, critical reflection, and independent decision making (Holec 1979; Keegan 1996; Little 1991). According to this view, the autonomous learner is capable of managing, monitoring and evaluating his or her own learning; and this capacity for self-regulation enables the learner to transcend the barriers that pedagogy often erects between learning and living (Barnes 1976). In the particular case of foreign languages, an autonomous learner is able to reflect critically both on the learning process and on the target language, and will continue to use this ability in contexts of target language use long after the end of formal learning. According to this definition, the freedom that learner autonomy confers begins as a matter of psychology rather than organisation. As a pedagogical goal autonomy is thus no less appropriate to classrooms than to self-access systems, no less important to children at primary school than to students at university.

This view of learner autonomy has gained widespread recognition in recent years, as more and more national and regional curricula have come to emphasise the necessity of developing learners' capacity for critical reflection and independent thought. And yet, even at university level, fully achieved learner autonomy remains a rarity. Perhaps the chief reason for this lies in a failure to recognise and respond to the paradox on which the development of learner autonomy is founded: in order to achieve autonomy, learners must first be autonomous. We can see that this is a paradox and not a contradiction if we consider the case of developmental learning, an enterprise in which all normally endowed children succeed.

Autonomy, in a broad behavioural sense, is clearly the *goal* of developmental learning: we count as mature only when we are able to act for ourselves across a wide range of social contexts. But autonomy is also the *mode* of developmental learning: environment plays an important role in fuelling and to some degree shaping development, but from the beginning the initiative lies with the individual child. In this sense we are self-producing organisms (cf. the notion of "autopoiesis" elaborated by Maturana & Varela 1992). Clearly, a toddler of 3 is capable of a much narrower range of independent behaviour than a child of 10; and a child of 10 is in turn capable of a much narrower range of independent behaviour than an adolescent of 15. But each behaves autonomously to the limits of his or her capacity; and at every level developed autonomy provides the necessary precondition and indispensable basis for further development.

To begin with, the autonomy that is at once the mode and the goal of developmental learning is an implicit, unconscious phenomenon; and the extent to which it becomes explicit (that is, something the individual can consciously reflect on and talk about) is infinitely variable (cf. Dworkin 1988). By contrast, the autonomy that is the goal of formal learning is always an explicit phenomenon, since it is founded on such conscious, intentional activities as critical reflection and evaluation. Nevertheless, in formal no less than in developmental learning, the growth of autonomy from one level to the next presupposes a pedagogy that from the first requires learners to be autonomous, to accept responsibility for their own learning, to the extent that their current capacity allows.

Autonomy develops through interaction, and the independence that it confers is inseparable from the interdependence that frames its growth. In developmental learning, the child's inborn capacity for autonomy grows under the stimulus of interaction with the environment; children kept in isolation or under severe restraint simply fail to develop in what we recognise as a normal way (cf. the case of Genie; Rymer 1994). Similarly in formal learning, the capacity for autonomy develops from appropriately focused interaction with teachers and other learners. When learners are not involved in such interaction, they will not usually develop a capacity to manage their own learning.

The argument so far has two implications for self-access language learning systems. First, learners who have experienced a pedagogy oriented to the development of learner autonomy are best placed to take advantage of the learning opportunities offered by such systems, since they know what it is to be responsible for one's own learning – to select content, plan and monitor activities, and evaluate progress. Second, as a consequence of the fact that learner autonomy is not a common reality in our universities, self-access systems must themselves help learners to become self-managing and self-regulating. One way of doing this is by providing some kind of advisory service; another way is by offering possibilities of interaction with other learners. Tandem learning is one of the forms that the latter kind of interaction can take; and as we shall see, tandem learning poses a sharply focused challenge to each partner's capacity for autonomous behaviour.

2. Tandem language learning

In tandem language learning two people with different mother tongues work together in order to learn each other's language, learn more about each other's character and culture, and perhaps exchange additional knowledge – e.g., about their professional life (Little & Brammerts 1996:10). In its original form, tandem language learning is conducted face-to-face; the partners meet regularly at an agreed time, and their primary channel of communication is oral. It is important to emphasise that a successful tandem partnership is not simply a convenient way of getting conversation practice: it is explicitly a *learning* partnership. Each meeting should be devoted to the collaborative performance of agreed learning activities, and each partner should be intent on helping the other to identify and find solutions to learning problems.

Tandem language learning is governed by the principles of reciprocity and learner autonomy (Little & Brammerts 1996:11). The principle of reciprocity requires that both partners contribute equally to their collaboration; while the principle of learner autonomy requires that partners are responsible for their own learning. Because tandem learning is a partnership and learner autonomy develops interactively, accepting responsibility for one's own learning necessarily entails accepting responsibility also for the learning of one's partner.

Tandem language learning can be organised in various ways. For example, it may be an integral part of a larger course of language learning, an optional supplement to a language course, or the only form of language learning in which the partners are engaged. Each of these frameworks may exist in association with a self-access system and its advisory service, though each makes different demands of the tandem partners. When tandem learning is an integral part of two language courses that share the same aims, content and structure, organisation is principally a matter of the partners agreeing how often they should meet, how long their meetings should last, and where those meetings should take place. The learning activities they engage in will be, at least to some degree, determined by their respective classes, which may be expected to devote time to advising learners how to derive maximum benefit from their tandem partnership. When tandem learning is an optional supplement to a language course, on the other hand, and especially when it is the only form of language learning they are engaged in, the partners will probably have to select learning activities for themselves. In order to do this they must have a clear view not only of each other's short- and long-term learning goals but also of the processes involved in language learning. Unless they are already far advanced in the ways of learner autonomy, they are unlikely to make much progress in their partnership without advice and support from an expert. Precisely because tandem learning is founded on partnership, it is desirable that advice should be offered to both partners together rather than individually; and because the long-term success of each tandem partnership depends on the development of both partners' autonomy as learners, advice should not be handed down in the form of rules to be obeyed, but mediated through exploration, reflection and negotiation.

The principle of reciprocity requires that each meeting between tandem partners should be bilingual, half the time being devoted to one language and half to the other. This means that in each meeting both partners are cast in two roles, language learner and native speaker. The benefit that accrues to tandem partners as language learners is that they have the opportunity to communicate regularly with a native speaker of their target language, and to do so while maintaining a focus on their present learning concerns. In order to maximise this benefit, it is essential that it is always the learner who decides the direction and shape of the interaction by taking most of the discourse initiatives; for in this way his or her learning remains the centre of concern and the native speaker cannot slip into the role of (untrained) teacher. This is not to say, of course, that the learner should not explicitly seek feedback from the native speaker; on the contrary. But when feedback is sought by the learner rather than volunteered by the native speaker, it should have to do with aspects of the target language the learner wants to hear about rather than aspects the native speaker (rightly or wrongly) considers important.

Conversation between a small child and her mother provides a good general model for tandem learning encounters. The child as learner typically takes the initiatives that determine the shape of the conversation, while the mother provides the discursive support necessary to keep the conversation going (she clarifies, rephrases, expands, and comments), and in this way the child's communicative capacity gradually grows. Note that in the course of such a conversation it is by no means unusual for the child to ask a question that is explicitly linguistic, focusing perhaps on the meaning of a word or its pronunciation or grammatical form. Very often, of course, child–mother conversations are built around some shared activity, like building a house with wooden bricks or giving a dolls' tea party, just as conversations between tandem partners are built around the collaborative performance of a learning task.

At first sight it might appear that all the benefits in a tandem partnership accrue to the learner; that playing the role of native speaker for half of each encounter is simply the price one pays for spending the other half communicating in one's target language. But this is a serious misapprehension. In a successful tandem partnership each partner can in principle benefit as much from the native speaker as from the learner role. For provided the learner takes the decisive discourse initiatives, the native speaker has an unparalleled opportunity to experience and reflect on his or her mother tongue through the prism of the target language and vice versa. For example, the errors that the learner makes will in many cases reveal important contrasts between the two languages and in this way throw light on the native speaker's target language problems.

If their previous language learning experience has been shaped by a traditional teacher-led pedagogy, tandem partners will not find it easy to enact the roles of learner and native speaker as I have described them. In traditional pedagogy the initiative lies with the teacher rather than the learner, and this may lead the partners to suppose that in each half of their encounters the native speaker should take the dominant role. This, no less than the matters of organisation referred to above, is an issue on which the great majority of tandem learners will need advice and support. Again, it is desirable that advice should be mediated via exploration, reflection and negotiation with both partners together. Almost certainly it is not enough simply to describe how the partners should enact the roles of learner and native speaker: the adviser needs to model them with each partner in turn.

As a tandem partnership develops and the two partners become more skilled in the performance of their dual roles, so their autonomy grows as regards both language learning and target language use. It is fundamental to the nature of tandem learning that this growth is the product of an interdependent, collaborative process (compare with this my earlier argument concerning the growth of autonomy in developmental learning). Especially at the beginning of their partnership tandem learners are likely, as I have argued, to require advice and support from an expert. But although this may play a crucial role in helping them to understand how to structure their encounters and support their own and their partner's learning, their development as learners and users of their respective target languages is driven by the reciprocal dynamic of their partnership. This is the unique power of tandem learning, which deserves a great deal more theoretical and empirical exploration than it has so far received.

3. Tandem language learning via the Internet

Perhaps the best way of exploiting, but also systematically exploring, the power of face-to-face tandem language learning is to organise two language courses at the same time and in the same place – say, English for native speakers of German and German for native speakers of English – and to bring the participants in both courses together in tandem partnerships. The two groups of learners should be broadly comparable in their target-language proficiency, and the two courses should be as far as possible identical in their aims, structure and pedagogical approach. Ideally, the teachers of the two courses should be fluent in both English and German, so that they can collaborate flexibly in advising and supporting tandem partnerships. Such an arrangement is likely to require that either the native speakers of German travel to England or Ireland, or the native speakers of English travel to Germany. In other words, it is likely to be possible only on a short-term basis – as a residential summer school, for example.

In all other circumstances, face-to-face tandem partnerships are difficult to establish and maintain, even on a modest scale. In British and Irish universities, for example, the number of students learning French at any one time will always be many times greater than the number of native French speakers who are enrolled as visiting or exchange students – and in any case, by no means all of these native French speakers will be interested in forming a tandem partnership. The International E-Mail Tandem Network, which is co-ordinated by Helmut Brammerts at the Ruhr-Universität Bochum (Little & Brammerts 1996; http://www.slf.ruhr-uni-bochum.de), is a response to this problem. Via a series of bilingual sub-nets, it provides a means of putting Spanish-speaking learners of German in touch with German-speaking learners of Spanish, English-speaking learners of French in touch with French-speaking learners of English, and so on, without requiring them to meet face-to-face. Each sub-net is supported by a bilingual discussion forum, and individual learners are assigned tandem partners by a "dating agency" in Bochum. As its name implies, the International E-Mail Tandem Network at first made use exclusively of e-mail as the channel of communication between learners. More recently it has also begun to exploit the synchronous text-based communication that is available in MOOs (*M*ulti-user domains, *O*bject-*O*riented; originally developed to support role-playing games on the Internet). For this purpose Klaus Schwienhorst of the Centre for Language and Communication Studies, Trinity College, Dublin, developed a virtual language centre at Diversity University (cf. Schwienhorst 1997; http://www.tcd.ie/CLCS).

The existence of the International E-Mail Tandem Network does not in itself guarantee the success of tandem language learning via e-mail, as members of the network have been quick to discover. For one thing, although e-mail is by now established as a preferred channel of communication within and between universities around the world, the extent to which individual students have access to e-mail is very variable. Thus one member of a tandem partnership may be able to check her e-mail several times a day, while the other is lucky if she can do so twice a

/eek. In cases like this, tandem partners will have very different views of how
ften they should write to each other, and they may never achieve a clear under-
tanding of the very different situations in which they are placed. A second problem
s that some university computer networks are better specified and more reliable
han others. Here too disproportion between institutions can have a powerful nega-
ive impact on tandem partnerships. If the system in one's own university is often
down", or one's partner seems able to send messages only with the greatest diffi-
ulty, tandem language learning by e-mail will quickly be perceived as not worth
he effort. A third practical problem arises from differences in term or semester
tructure from one country to another. It is not simply that, for example, students in
reland and Germany are on vacation at different times; their academic years follow
quite different rhythms, so that they are preoccupied with examinations and other
listractions from tandem learning at different times. Taken together these factors
nean that Irish–German tandem partnerships can be actively pursued for only about
welve weeks in each academic year.

Besides these general organisational problems, tandem language learning via e-
mail also has to overcome problems arising from the fact that partners collaborate
at distance rather than face-to-face, by writing rather than by speech. On the one
hand it is more difficult to establish an effective working relationship at distance
than face-to-face, not least because the reciprocal nature of face-to-face communication allows partners quickly to form an intuitive sense of what will and what will
not help their relationship to develop. On the other hand, it is easier to opt out of an
e-mail partnership than one that is conducted face-to-face, because an e-mail partnership is largely free of the complex social pressures that a face-to-face partnership quickly generates. In principle it may be possible to counteract these effects
through the use of MOOs for at least some of the communication between tandem
partners. It is important to note, however, that precisely because they involve synchronous communication, MOOs give rise to new organisational problems; also
that many universities cannot yet provide students with the necessary level of Internet
access.

Providing e-mail tandem partners with support is also not without problems.
The International E-Mail Tandem Network offers users on-line advice together with
suggestions for learning activities at its website in Bochum; and each sub-net has a
discussion forum where, at least in principle, learning problems can be reported
and possible solutions proposed. But if autonomy in tandem language learning as in
other contexts develops out of interaction, the fact that these supports are delivered
as written text must constrain their effectiveness: whereas face-to-face oral communication is a reciprocal activity to which all participants can contribute simultaneously, written communication is non-reciprocal in the sense that writers and readers
must construct and understand messages without immediate interactive support from
the message receiver or producer. Of course, self-access systems and other institutional structures can provide a face-to-face advisory service for their half of each
tandem partnership. But although much may be achieved in this way, it is a serious
disadvantage not to be able to advise tandem partners together. MOOs offer a means

of overcoming this limitation (the virtual language centre at Diversity University has a room devoted to tandem counselling), but advising in virtual reality has yet to be implemented and evaluated on a systematic basis.

Because of these problems, and bearing in mind that most students have a seriously under-developed capacity for autonomous learning, it seems likely that tandem language learning via the Internet will work best when, on both sides of the partnership, it is fully integrated in a larger programme of language learning. This requires close collaboration between two institutions offering closely similar courses. Trinity College Dublin and the Ruhr-Universität Bochum undertook just such a collaboration as part of the EU-funded project "Telematics for Autonomous and Intercultural Tandem Learning", which was co-ordinated from Bochum (1996–8). The collaboration entailed the establishment of tandem partnerships between students learning German in Dublin and students learning English in Bochum. On both sides of the collaboration the students were following extra-curricular language courses that sought to promote language learning by language use, emphasised the importance of learner autonomy, and were based on project work. Every effort was made to overcome the organisational problems outlined above. For example, because of the asymmetry between the Irish and German academic years, tandem learning was an integral part of only two out of four project cycles; to begin with, students were "double-dated" to maximise their chances of establishing a workable tandem partnership; and the scheme was supported by its own discussion forum. Students on both sides were obliged to engage in tandem learning and to submit a minimum of four tandem exchanges as part of the dossier of work that underpinned the preparation of the two projects in question. The virtual language centre at Diversity University was used for a limited number of tandem encounters, including a certain amount of counselling. In theory the development of learner autonomy within tandem partnerships should have been greatly assisted by the fact that the courses from which they arose were shaped by a pedagogy in which learner autonomy played a central role: by the time students embarked on tandem learning they were used to collaborative working methods that were shaped by ongoing negotiation and self-evaluation. (An evaluation of the Dublin–Bochum project is provided by Little et al. 1999.)

In order to derive maximum benefit from tandem learning via the Internet it is essential to be clear that it differs from face-to-face tandem learning in important respects. In face-to-face tandems, the primary channel of communication is oral and reciprocal. This means that partners negotiate meaning interactively, which in turn means that in every exchange the native speaker can help the learner to get his or her message across (cf. the example of child–mother interaction cited above). Tandem learning via the Internet, by contrast, is text-based and thus non-reciprocal in the sense defined above: each message must be produced by one of the partners working alone; when he is communicating in his target language he gets no immediate interactive support from his native speaker partner. Tandem communication by e-mail thus requires a greater capacity for autonomous language use than face-to-face tandem communication; many learners will need help in formulating their messages, at least in the early stages.

At the same time, however, use of the written channel brings two notable advantages to counterbalance this disadvantage. The first of these arises from the permanence of writing compared with the transience of speech. Face-to-face tandem partners may agree to record their meetings on audio-cassette, but this will benefit them only if they are prepared to spend a lot of additional time listening and perhaps transcribing. In most cases, their meetings will survive only in the written notes they make. A tandem partnership conducted by e-mail, on the other hand, can be preserved in its entirety. It can be printed out, analysed, shared with other learners, filed, and re-read months after the event. In other words, an e-mail tandem partnership not only supports learning now, but provides a resource for learning in the future. The same is true of partnerships conducted in text-based virtual reality.

The second advantage that use of the written channel brings has to do with the development of learners' metalinguistic awareness, which is essential for reflection on language learning and language use, and thus a precondition of learner autonomy. While it is true that we can focus on matters of linguistic form while we are communicating orally, it is also true that we find it easier to do so by reference to a written text, because writing makes language visible and at the same time takes it "off line". This means that tandem learning by e-mail should have a greater capacity to foster the development of metalinguistic awareness than face-to-face tandem partnerships. By the same token it should also encourage partners to fulfil their native speaker role of providing feedback in an appropriately reflective manner.

4. Conclusion: Future prospects

Tandem language learning is one of the most powerful means by which self-access systems can help students to develop their capacity for autonomy both as language learners and as users of their target language. In principle, e-mail and MOOs make tandem partnerships a possibility for very large numbers of students around the world; though as we have seen, the implementation of tandem learning in this domain is vulnerable to a number of potentially disabling organisational problems. Even when two institutions collaborate closely to overcome these problems and create conditions that are maximally favourable to tandem learning via the Internet, pedagogical problems are bound to arise. It is possible to predict the general nature of these problems from tandem principles on the one hand and fundamental characteristics of written communication on the other. But we cannot do more than that until carefully controlled experiments have been subjected to empirical exploration and evaluation. Such exploration and evaluation will not only help us to refine our organisational frameworks and pedagogical procedures; it will also increase our understanding of tandem learning and the principles of autonomy and reciprocity on which it is founded. In this way it will help to prepare us for the new modes of tandem learning that future developments in Internet communication will bring.

References

Barnes, D. (1976). *From communication to curriculum*. Harmondsworth: Penguin.

Dworkin, G. (1988). *The theory and practice of autonomy*. Cambridge: Cambridge University Press.

Holec, H. (1979). *Autonomy and foreign language learning*. Strasbourg: Council of Europe.

Keegan, D. (1996). *Foundations of distance education* (3rd ed.). London & New York: Routledge.

Little, D. (1991). *Learner autonomy 1: Definitions, issues and problems*. Dublin: Authentik.

Little, D. (1997). Autonomy and self-access in second language learning: Some fundamental issues in theory and practice. In M. Müller-Verweyen (Ed.), *Neues Lernen – Selbstgesteuert – Autonom / New developments in foreign language learning – self-management – autonomy* (pp. 33– 44). Munich: Goethe-Institut.

Little, D. & Brammerts, H. (Eds) (1996). *A guide to language learning in tandem via the Internet*. CLCS Occasional Paper No. 46. Dublin: Trinity College, Centre for Language and Communication Studies.

Little, D., Ushioda, E., Appel, M.C. Moran, J., O'Rourke, B. & Schwienhorst, K. (1999). *Evaluating tandem language learning by e-mail: Report on a bilateral project*. CLCS Occasional Paper No. 55. Dublin: Trinity College, Centre for Language and Communication Studies.

Maturana, H.R. & Varela, F.J. (1992). *The tree of knowledge. The biological roots of human understanding*. Boston & London: Shambhala.

Rymer, R. (1994). *Genie: A scientific tragedy*. Harmondsworth: Penguin.

Schwienhorst, K. (1997). Virtual environments and synchronous communication: Collaborative language learning in object-oriented multiple-user domains (MOOs). In D. Little & Voss, B. (Eds), *Language centres: Planning for the new millennium* (pp. 126–144). Plymouth: CERCLES.

3

Learner autonomy, self-instruction and new technologies in language learning: Current theory and practice in higher education in Europe

Jeannette Littlemore
University of Birmingham, UK

1. Introduction

A central goal of current approaches to language teaching including communicative language teaching, task-based learning and learning strategy training is to enhance student autonomy and control over the language learning process. This has been particularly true of language teaching in Higher Education (HE). Two concepts which have thus become important in language teaching are learner autonomy and self-directed learning, concepts which are discussed below. It has been argued (Pennington 1996; Warschauer et al. 1996) that new technologies (computers, the Internet, multimedia and so on), as well as some not-so-new technologies (language laboratories and video) can assist in the development of learner autonomy and self-directed learning. Indeed, with the emergence of more and more self-study centres and the increasing enthusiasm with which HE language departments embrace new multimedia, the relationship between technology and independent learning needs to be studied, especially as this relationship may not be as straightforward as it seems. This chapter seeks to compare the prevalent theories with current practice, examining ways in which new technologies can be used and are being used to promote learner autonomy and self-directed language learning in HE institutions across Europe.

This study is divided into two parts. In the first part, current theories are presented concerning the relationships between learner autonomy, self-directed learning and new technologies. It is pointed out that, although new technologies can indeed be used to promote different types of independent learning, careful reflection is required if they are to be used effectively. The second part of the chapter presents findings from a survey which consisted of a questionnaire and interviews with 40 HE language-teaching institutions (mostly universities) in Europe. These findings have made it possible to assess the extent to which current practice in the area reflects current theories. The conclusion outlines several avenues which might be explored in order to improve and extend both knowledge and use of learner autonomy, self-directed learning and new technologies.

2. Theories about learner autonomy and self-directed learning, and ways in which they can be supported by new technologies

The rapid evolution of this area of research means that definitions of autonomy change as new theories are put forward and new practices are piloted. Although the meaning and implications of autonomy have become clearer over the last twenty years, there is still some confusion about the whole issue. Therefore this chapter seeks to establish exactly what is meant by the terms "learner autonomy" and "self-directed learning". These definitions will serve as the basis for the subsequent examination of some of the ways in which new technologies can support learner autonomy and self-directed learning.

In order to understand the concept of learner autonomy, it is perhaps easiest to state what learner autonomy is *not*. As Little (1991) points out, where there is hostility towards the idea of learner autonomy it is often based on a false assumption about what autonomy is and what it entails. He outlines five common misconceptions about autonomy:

- A first misconception is that autonomy is synonymous with self-instruction, that it means simply working without a teacher. It is certainly true that some learners who follow the path of self-instruction achieve some degree of learner autonomy, but many do not.
- A second misconception is that in order to encourage autonomy in the students, the teacher must relinquish all control in the classroom, as any intervention on the part of the teacher may destroy whatever autonomy the learners have managed to attain. This is simply not true; autonomy is still possible in a class where the teacher remains in control.
- A third misconception is that "learner autonomy" is a new methodology, that it is something which can be programmed into a series of lesson plans. Unfortunately the concept of autonomy is not so simple.
- A fourth misconception is that autonomy is a single, easily described behaviour. This is also false. Autonomy can, according to Little, take many different forms, depending on the age of the learners, their stage of learning and their learning goals, etc.
- A fifth misconception is that autonomy is a steady state achieved by certain learners. In reality, the permanence of autonomy cannot be guaranteed, and a

learner who displays a high degree of autonomy in one area may not be autonomous in another. (Little 1991:3–4)

Little's definition of autonomy is as follows:

> Essentially, autonomy is a capacity – for detachment, critical reflection, decision-making, and independent action. It presupposes, but also entails that the learner will develop a particular kind of psychological relation to the process and content of his learning. (Little 1991:4)

The main advantage of learner autonomy is, according to Little, that an agenda set by the learner himself is likely to be more purposeful and effective both immediately and in the longer term. He also makes the point that barriers between learning and living are lifted in autonomous learning. Both of these points are crucial to language learning, where the issue of relevance is central. Learners of modern languages usually intend to use the language which they are learning, usually to express concerns which are of importance to them.

Holec (1988) discusses the relationship between the concepts of learner autonomy and self-directed learning. He describes autonomy as "the ability to take charge of one's own learning". This ability is, according to Holec, not something we are born with; it has to be learnt either consciously or unconsciously. When a learner has acquired this ability he/she must have the chance to use it and be willing to use it. Ellis & Sinclair (1989:2) also emphasise the importance of learner training in the development of autonomy, claiming that it can enable learners to "make informed choices about what, how, when and where they learn". The learner who has this ability and makes full use of it can be said to be engaged in *self-directed* learning. Holec points out that "self-directed learning implies an autonomous learner" (1988:9). On the other hand, an autonomous learner need not use his/her ability completely and may be only partially involved in self-directed learning. Holec explains the terminology in the following way: Autonomy refers to an ability, thus the adjective autonomous should only be applied to a person, not a process. This is why he suggests the expression "self-directed learning" to describe the learning process in which an autonomous learner is involved.

According to Holec (1988) there are varying degrees of self-direction in learning which may be connected to varying degrees of autonomy. He outlines the following components of an entirely self-directed process of learning:

- fixing the objectives
- defining the contents and progressions
- selecting the methods and techniques to be used
- monitoring the acquisition procedure
- evaluating what has been acquired

The main advantage of self-directed language learning is thus that learners can focus on the aspects of the target language that they consider to be of most use to them. They can also choose methods of learning which are best suited to their particular learning styles and personalities. This chapter uses the terms learner autonomy and self-directed learning in a way which incorporates the work of both Little and Holec.

It has been argued that new technologies, in particular computer networks, have the potential to increase learner autonomy when they are used appropriately. Warschauer et al. (1996) claim that:

> the mechanics alone of computer-mediated communication provide students with a much better opportunity for control and initiative in language learning. (Warschauer et al. 1996:3)

They give a number of examples including the following:

- The "asychronicity" of e-mail frees students from time and distance limitations, enabling them to initiate discussions with their teachers or with other students at any time of day and at a number of places rather than only during class or office hours.
- Writing instructors report that the use of computer conferencing can prompt more discussion that is student-centred.
- When long distance communication is available, students have the independent opportunity to use the foreign or second language for authentic communication with native speakers.
- Many teachers suggest that linking cross-cultural communication through e-mail with task-based learning provides the most fruitful combination for fostering student autonomy. (Warschauer et al. 1996:3–5)

Finally, in relation to learner autonomy, Warschauer et al. claim that computer networking can develop students' learning skills and critical learning perspective. One of the main supporting arguments for this claim is that, with the "information explosion", knowing how to find and interpret facts is more important than memorising them.

Despite the optimism of writers such as these, the relationship between learner autonomy and new technologies is not straightforward. It has been pointed out that the increasing opportunities offered by new technologies do not necessarily make these technologies automatically favourable to learner autonomy:

> while it is quite clear that improved and immediate access to an ever wider range of pedagogical and non-pedagogical language resources cannot be other than beneficial to self-instruction, the potential impact on autonomy is a more complex issue. (Kenning 1996:122)

In support of her argument, Kenning outlines four types of independence which constitute learner autonomy. These are *physical independence* (learning on one's own without a teacher), *social independence* (not relying on a teacher for the provision of social interaction), *linguistic independence*, (the use of the target language to access other knowledge domains), and *managing one's own learning* (organising both the cognitive and the meta-cognitive aspects of learning). She claims that computer-assisted language learning can help the first three types of independence, but that *the management of one's own learning* is not always enhanced by computers, particularly when using types of software where a central and directive role is taken by the computer. Another way in which, according to Kenning, didactic software can impede learners' ability to manage their own learning is by providing instant feedback. This may lead to machine dependency, as it denies the students the opportunity to reach decisions on their own, and to develop their own discovery procedures. Even the computer when used as a tool (for example the use of word processing packages equipped with spell checks and grammar checks) can, according to Kenning, lead to learner dependency if students are not encouraged to use appropriate strategies within an overall learning framework. Students need to be trained in the use of these strategies, as they cannot be expected to adopt them automatically. Finally, Kenning points out that the use of IT in language learning increases the students' need for information-handling skills. She maintains that these skills do not come naturally and that students need to be trained in determining objectives, matching activities to the learning process, and in diagnosing their own difficulties. As we shall see, this emphasis on strategy training is particularly important in the context of this chapter.

Thus, in conclusion to this first section, the predominant opinions held by researchers into learner autonomy, self-directed learning and new technologies may be summarised as follows:

- Learner autonomy is distinct from self-study in that it implies a capacity for detachment enabling the learner to decide for him/herself how, why, when and where to learn.
- Self-directed learning refers to the learning process in which an autonomous learner is involved.
- New technologies can be used to encourage different types of independent learning but do not automatically do so; care must be taken not to replace "teacher dependency" with "machine dependency".
- Learners need to be trained in the strategies required to make the most of the opportunities offered by the new technologies.
- It is important that learners continue to have support from their teachers. They must not simply be left alone with the new technologies.

3. The relationship between theory and practice: Attitudes towards learner autonomy, self-directed learning and new technologies in Higher Education language teaching in Europe

This section is based on findings from a survey which was carried out in 1997 into the use of new technologies to support learner autonomy in 40 HE language teaching institutions throughout Europe. By means of questionnaires and in-depth interviews, directors of university language centres were asked to relate their experiences with learner autonomy and new technologies. The aim was firstly to identify the extent to which the various departments differentiate between learner autonomy and self-instruction, and the degree to which these are thought to be relevant to their own teaching situations. A second aim was to establish the extent to which universities are able to provide the kinds of human support required by their students in order to become autonomous. Finally, the questionnaires and interviews aimed to discover how the roles of teachers have been changed by increased learner autonomy and self-instruction.

The survey revealed a wide range of attitudes towards learner autonomy, self-directed learning and new technologies. There was also wide variation in the universities' interpretations of the relationships between the three concepts. However, a strong correlation was observed between interest in the area and familiarity with current theories. Equally, where there was a lack of interest, this was often accompanied by a misunderstanding of the concepts, probably reflecting a lack of familiarity with current research. This correlation made it possible to group the universities into three types, each having its own characteristic attitude towards learner autonomy, self-directed learning and new technologies. These are outlined below.

3.1 Universities whose approach to learner autonomy and new technologies does not correspond to current theories

Just over a fifth[1] of the universities surveyed do not feel that the notion of learner autonomy is particularly relevant to their own particular situations. However, in many cases, these universities do not appear to be aware of the full meaning of learner autonomy, equating it instead with self-instruction. This view is exemplified by the comment that:

> [Learner autonomy] runs partly counter to the present view of efficient language learning and the communicative curriculum, except in specific areas, e.g. vocabulary learning. Self-instruction seems possible in other fields of English studies, e.g. specific areas in linguistics, literature where facts have to be learned.

Often, learner autonomy is only seen as being relevant where facts have to be learned. Any other kind of learning, apparently, requires a teacher. Learner autonomy is seen as a good way of cutting costs and teaching time. Large class sizes and limited time allocated to language teaching are considered to be acceptable reasons for promoting learner autonomy. While these may be very practical reasons, they do not seem to take into account learner autonomy as anything other than students learning on their own. Learner autonomy is often seen by these universities as something very practical, unrelated to research interests. It is thought that increased time spent creating materials for self-study would take up precious research time. Self-instruction is "offered", or "sold as supportive of instruction". It is not integrated into the main teaching schedule. Indeed the role of the teacher in these institutions remains very traditional. It is often the case that teachers already have very heavy teaching loads, and simply do not have the time to create materials for self-access centres. Language teachers are often found to be generally unenthusiastic about self-study and learner autonomy.

There is often little or no human support for independent learning in these universities. Once the students have established the framework of their particular project with the help of a supervisor, they are often left virtually alone with the study centre materials with the only assistance available being help with the hardware and software. This is usually for financial reasons or because self-study facilities are so underdeveloped that they have not been deemed worthy of support personnel. Lack of funding for study centre personnel is often a serious problem.

At one university, where very few language teachers have shown any interest in the use of new technologies, it is suggested that the root of the problem lies with the lack of contact and co-operation between linguists and computer science experts. It seems that this arts/science divide is an obstacle which will have to be overcome if the opportunities offered by new technologies are to be integrated into language learning. A possible solution might be the provision of specific funds and career opportunities for those willing to bridge this gap.

Thus it can be said that the views of the universities in this category towards learner autonomy, self-directed learning and new technologies do not correspond to current theories in the area. This is often because they do not see the concepts as being particularly relevant to their needs. However, this can be seen as a "chicken and egg" issue. If current theories in the area were more widely disseminated, then universities such as these might see the relevance of such concepts as learner autonomy and self-directed learning. Exposure to the wide range of opportunities offered by new technologies might allow them to perceive a more useful role in their language teaching provision.

3.2 Universities whose approach towards learner autonomy and new technologies corresponds in part to current theories in the area

Over half the universities surveyed have self-study facilities and use new technologies, but do not have a vision of the long-term effects that these will have on their language teaching provision. Self-study facilities are available but are not well incorporated into the general curriculum, and in some cases difficulties have been encountered in motivating students to use the facilities. Often, the approach to learner autonomy is very much technology-driven. New computers arrive and departments are faced with the question of what to do with them. This often engenders a degree of staff scepticism, which is difficult to overcome.

Although very much in favour of learner autonomy, some universities find it difficult to motivate their students to become autonomous, even when the materials are highly relevant to their needs. There is often very low student turnout for self-study centres and there are frequent claims that students still seem to need teachers present, and are not willing to explore reasons for their errors independently, turning instead to the teacher for an explanation. The teacher constantly needs to act as a motivator and facilitator, spurring students to make full use of all that is on offer within a piece of software. Although learner autonomy and self-directed learning are sometimes seen by management primarily as both a cost-cutting device and a natural result of developments in computing technology, the teaching staff occasionally manage to benefit from this, and conduct research into ways of increasing learner responsibility. It seems that these teachers are managing to put financially-motivated decisions to practical pedagogical use.

Usually the role of the teacher is expected to remain that of an "imparter of information", but he/she is increasingly being asked to play a much more active and participatory role, providing guidance as to where to find additional information. In some cases he/she has to prepare self-study materials for the students. Although teachers are spending more time designing, implementing and administering self-instruction schemes, their central duty is still teaching. At some universities a few members of the language teaching staff have modified their teaching in order to increase learner autonomy (for example, by starting to provide counselling while relying less on traditional teaching in their seminars), but this only applies to the minority of department staff. The syllabi and curricula used by many universities are, apparently, still too rigid to allow any real learner autonomy. Furthermore, learner autonomy is reported as being "mistakenly thought of as being some sort of 'laissez faire' policy by a number of teachers", and there is, according to one response, a need for further research and discussion of the topic, for example by means of staff symposia. It is often felt, in these universities, that the classroom and its teacher should still be at the centre of the learning process. The teacher's way of integrating self-instruction in the learning process determines its use. It is suggested by some respondents that more enthusiasm for learner autonomy would be brought about if teachers were given the opportunity to see successful working examples in other language teaching institutions. This is because there is thought to be a lack of awareness among teachers as to what exactly would be required of them by an increase in learner autonomy.

As far as human support is concerned, this is usually available, but often the students remain unsure as to the best ways of using the facilities for their independent learning, and the teachers are also left in the dark as to possibilities and limitations of the new materials. In a small number of universities the self-access centre is staffed by one full-time person and several part-time helpers but it is not clear whether these members of staff are language teachers, language advisers or technicians. Technical assistance is often available but the advisers are usually techni-

ians who double-up as tutors. Although this is going some way to recognising the
eed for human support, the technicians are not trained language teachers and as
uch will not be able to give advice on matters such as appropriate learning strate-
ies for learner autonomy.

Unfortunately, departments interested in the implementation of self-directed
earning have sometimes encountered the problem that the language department is
he only department within the university to encourage it. The fact that self-directed
earning does not correspond to the general learning culture of a university, or indeed of a country, would seem to be a problem encountered by many language
departments which have themselves embraced the idea and would like to implement it. As all other subjects are taught in a very traditional way, the students are
used to a highly structured "top down" learning culture. This means that the language department may have difficulty in encouraging its students to learn independently.

Thus it can be said that the universities within this, the largest, category are
going some way towards embracing the concepts of learner autonomy and self-
directed learning, with the help of the resources provided by computer-assisted
language learning. Their commitment to new technologies is, of course, limited by
both their financial resources and their existing teaching frameworks, but they see
an increasing role for them in the future. Their understanding of learner autonomy
and self-directed learning usually reflects current theories in the area. This is an
encouraging finding, as it suggests that those universities who are interested in
implementing these ideas are able to gain access to relevant research publications.

3.3 Universities where current theories on learner autonomy and new technologies have been fully embraced

The remaining universities (about a quarter) have a very clear idea of what learner
autonomy is. They have clearly defined objectives in this area. Learner autonomy
is strongly encouraged and self-access is fully integrated into the main language
curriculum. Universities such as these claim that:

> better learners (i.e. those who seem most often – though not always – to
> progress) are those who are aware of their own learning behaviour and are
> able to make independent choices about and evaluation of their learning.

Often there is a perceived need to teach language students "a way of continuing to
learn", placing great emphasis on the "creation of individuals who have acquired
the tools to become autonomous learners". It is not assumed that learners are going
to become autonomous by themselves. A need is perceived for preliminary orientation and further monitoring, and the teacher is seen as an important factor in helping the learner to become autonomous. Continuous self-assessment by the learners
is strongly encouraged. Reflective practices are seen as an essential part of all language studies and curricula are designed so that it is possible for teachers to embed

opportunities for learner autonomy in their courses, and give the students a real opportunity to build their own curricula throughout their studies.

A good example of how theories of learner autonomy and self-directed learning are put into practice is provided by one university where, after being given a brief introduction to how the system functions, the students are encouraged to work towards a personalised final goal according to the following procedure:

i   Together with their teacher they decide upon a suitable overall learning goal.
ii   They then decide upon smaller short-term learning goals that will go towards making up the overall learning goal.
iii They choose the material that they will require (books, cassettes, multimedia and so on) and the kinds of human support that they might require.
iv   They choose the kind of activity which will form the first learning goal.
v   They evaluate what they have learned.
vi   They revise what they have learned.

Self-directed learning is often viewed by these universities as a practical way of accommodating individual differences amongst students. After an initial entry test the teacher and student work together to produce a personal work plan based on the student's level in the target language, his/her immediate needs and preferred style of learning. The student then chooses the types of equipment and human support required to carry out this work plan. This flexibility is seen as an important factor in the decisions of a number of universities to promote self-directed learning for very practical reasons. One claims that many of its students are learning "with rather than at" the university, as many of them live abroad or study part-time and have to fit their education around their jobs. This is self-directed learning born of necessity, as well as responding to recent thinking in language teaching research.

As far as human support is concerned, all of the universities in this category see it as central to the successful running of their self-study centres. Their language teachers are confident with the technologies available, technicians ensure maintenance, and the university trains the teachers or employs full-time "language advisers" to help the students gain true autonomy. The role of the language adviser is to act as a bridging figure between the teacher, the learner and the new resources. The tasks and responsibilities of a language adviser usually involve:

• eliciting the learners' needs
• obtaining relevant information for the design of a study plan (e.g. educational background, learning styles and perceptions, time management, aims, etc.)
• providing adequate and clear guidance and support for learners to work autonomously
• monitoring the learning patterns of the users of the service and providing relevant and effective feedback
• helping the university to provide appropriate language learning opportunities

- monitoring resources in relation to users' needs
- training users to become proficient learners through better understanding of their learning processes
- acting as 'mirrors' and reminding learners of their original aims and objectives
- helping them to find and keep their motivation.

Users of a self-access centre will usually contact an adviser to start an individual, personalised programme in which they decide the pace, sequence and mode of learning as well as content and assessment criteria. The learner may need preliminary guidance from the advisor and further monitoring.

The human support in a self-study centre sometimes comes from the students themselves. The principles of learner independence and learner responsibility are taken to their natural conclusions when students participate in the development of the self-study facilities, not only in the early planning stage, but also later on when choosing the equipment and negotiating with computer firms. At one university, students have been involved in all decisions about what kinds of materials are needed and how the centre should best be organised. A student support team is in charge of running the self-study centre. This team makes decisions to purchase materials, conducts software training courses and organises department events. Individual students even offer workshops (poetry workshop, film workshop) or tutorials (essay writing tutorial), and develop materials for their peers (CALL exercises, program documentation and customised manuals). Furthermore, a large number of student volunteers supervise the centre during opening hours as the university cannot afford to pay for the permanent staffing of the centre.

Most universities claim that students need to be trained in language learning strategies and that this strategy training needs to be both systematic, explicit (during induction courses) and embedded (within the curriculum). It helps to reinforce the practice if training for independent study is provided centrally as well as locally. In order to motivate students to use self-study facilities, the outcomes of students' independent learning are sometimes assessed and accredited. Furthermore, "value-added" skills such as decision-making, time-management and technical ability are accredited. It is also recognised that teachers need support if they are to make good use of new technologies. At one university a multimedia manager coordinates and supports the development of materials by teachers, attuning teacher expectations and computer possibilities and limitations.

The most radical departments claim that the role of the teacher will change beyond recognition over the next few years. A wide variety of words is used to describe the new role of the teacher including "guide", "facilitator", "adviser", "enabler", "consultant", "organiser", "co-operator" and "creator" (of new materials). A few departments stand out as particularly forward-thinking in their expectations for the role of the teacher both now and in the future. For example, at one university, lecturers are considered to be working as part of an "integrated learning support team" (other areas of support being computer-assisted language learning, the Internet and so on). Teachers are seen as "facilitators of learning" rather than content deliverers and subject experts. They have become curriculum designers and producers of multimedia module study materials. They are strongly encouraged to use new technologies to add variety and flexibility to the learning process.

It is in this third group of universities where there seems to be the strongest link between theory and practice. These are universities where the idea of learner autonomy is an integral part of the overall teaching philosophy, rather than being something which is tacked onto the existing language teaching provision. As far as new technologies are concerned, these universities appear to have taken on board the ideas of theorists such as Kenning (1996), who maintains that a degree of on-going strategy training is necessary if learners are to use the new technologies effectively in order to develop true autonomy. The universities in this category are well informed about current theories of learner autonomy, self-directed learning and new technologies, and have extensive experience in the implementation of these theories. This suggests that they should share their knowledge and experience with universities which have been less successful, or which have not so far shown interest in promoting these concepts. This would give the latter a wider definition of the concepts, as well as providing them with opportunities to observe the benefits of self-directed learning in practice.

4. Conclusions

We have seen that there are many different attitudes existing towards learner autonomy across Europe and many different ways of implementing it. Universities at which it is not very well developed or supported tend to be those in which it is seen solely as a cost-cutting opportunity. Universities which support the idea, but where it does not seem to be working particularly well, should perhaps consider introducing compulsory self-study elements into their courses or offering students more guidance in self-study techniques and strategies. Where experiments in learner autonomy have enjoyed considerable success, this has happened because a great deal of effort has gone into helping the students set up their own work agendas, and assistance and advice have always been available throughout the learning process. The experience of these institutions makes it clear that learner autonomy does not mean simply leaving learners on their own to get on with it.

As far as human support is concerned, the most successful universities are those that have recognised the need for a substantial amount of support. This applies to both the learners (learning strategy training, introduction to new technologies and so on) and the teachers (training in how to use the new technologies, how to advise students, and so on). The amount of student support and advice seems to be one of the key determiners of a language department's success in encouraging learner autonomy. Several options are open when choosing what form this support should take. Teachers can be trained to act as advisers and consultants, special "language advisers" can be appointed whose job is purely to advise, or the students themselves can assume the role of support staff.

It seems that there are a variety of expectations for the future role of both the teacher and the learner. Some universities expect classroom learning to continue to form the core of language learning for years to come, while others expect a more radical change in the type of learning and the role of the teacher. There is of course room for all these possibilities, given the heterogeneous nature of university language departments across Europe, but it may be useful for a university to know where it stands in relation to others.

The first conclusion that can be drawn from this research is that there appear to be many universities whose practice in the use of new technologies to support learner autonomy and self-directed learning corresponds to contemporary theories of "best practice". A second conclusion that must be drawn, however, is that there is a great deal of variety, as the approach taken by universities necessarily must reflect their existing teaching ethos. Despite this diversity, four points have emerged which are common to the vast majority of universities and which merit further research. Firstly, increased use of new technologies and greater learner autonomy will require an investment of time. As is often mentioned in the replies to the questionnaire and in the interviews, more time will be required for training and for developing materials for self-study. Secondly the diffusion of information is very important. This includes theoretical information such as the meaning of learner autonomy as opposed to self-instruction, and practical information such as the kinds of human support required, ways of encouraging students to become autonomous and so on. Both types of knowledge seem to be unevenly spread between universities. Ways in which information can be transmitted from one university to another need to be created and extended. Thirdly, there is a marked gap between the high level technical skills and mastery of the new technologies shown by a small number of the centres, and the deficiencies in training and information shown by other centres. One of the challenges facing language educators will probably be to narrow this gap which exists both between and within countries. Finally, it seems likely that in the future different kinds of practices will co-exist, ranging from totally self-study to totally teacher-centred learning. It is unlikely that all HE establishments will promote learner autonomy and use new technologies in the same way, as different approaches will inevitably suit different learning contexts.

Acknowledgements

A large part of this article is based on information provided by language departments in universities across Europe. There is not enough space here to thank all the people involved in its preparation. However we must thank Judith Barna, Université Charles-de-Gaulle Lille III, France; Renate Faistauer, University of Vienna, Austria; Hamid Momtahan and Heather Matlock, Thames Valley University, England; and Richard Tuffs, The Open University, Belgium, for their particularly valuable contributions.

Note

¹   Given variations between the responses to the different questions, it is not possible to establish a perfectly clear distinction between the three categories. A very close approximation, however, would provide the following distribution:

Category 1: 7 institutions
Category 2: 22 institutions
Category 3: 11 institutions

References

Ellis, G. & Sinclair, B. (1989). *Learning to learn English*. Cambridge: Cambridge University Press.

Holec, H. (1988). *Autonomy – self-directed learning: Present fields of application*. Strasbourg: Council of Europe.

Kenning, M.M. (1996). IT and autonomy. In E. Broady & Kenning, M.M. (Eds), *Promoting learner autonomy in university language teaching* (pp.121–138). London: CILT.

Little, D. (1991). *Learner autonomy 1: Definitions, issues and problems*. Dublin: Authentik.

Pennington, M.C. (1996). *The power of CALL*. Houston, CA: Athelstan.

Warschauer, M., Turbee, L. & Roberts, B. (1996). Computer learning networks and student empowerment. *System, 24.1*, 1–14.

4

Criteria for the evaluation of Authoring Tools in language education

David Bickerton
University of Plymouth, UK
Tony Stenton
Université des Sciences Sociales, Toulouse I, FR
Martina Temmerman
Universiteit Antwerpen, UFSIA, BE

1. Introduction

Evaluation literature on Authoring Tools is replete with prescriptive and categorical assertions. Thus, it is claimed that they should be as easy to use as a text processor (Brücher 1993), or can be used by anyone (Locatis et al. 1991), or alternatively that their users properly require a background in programming (Strauss 1995). The polarisation implicit in such viewpoints is rendered explicit in product literature, as might be expected, and it is most certainly present in the mind of newcomers to Multimedia CALL authoring, who invariably ask: "Which is the best authoring tool?"

There is no longer much to be gained from such global characterisations. The truth is that whilst different products do suit quite distinct user groups, many authors are called upon to employ a variety of products and will shift periodically from one to another. The market has diversified both in terms of its products and its users, and diversification makes for plurality.

Products range from fully integrated systems with a generic educational profile to those targeting specialist application areas or providing components for other applications. Their functionality, which reflects the greater complexity of networked multimedia PCs, has also grown enormously in the last 10 years. Competition for distinctiveness means that Authoring Tools (ATs) are constantly upgraded or extended. In the brief time that has elapsed since the RAPIDO Project reported early in 1997[1], several "Internet-ready" releases of major ATs have been noted, together with a raft of WWW-delivered solutions. Current variations on HTML promise an even greater revolution in multimedia authoring.[2] It is clear, as a consequence, that techniques for evaluating ATs need to be kept under constant review and that "the decision-making process [...] is taking on new criteria" (Botto 1996).

The data from AT evaluations, ranking products according to specified criteria, are naturally subject to rapid flux given the speed of change of the educational IT industry. This chapter will re-assess techniques of AT evaluation from the broad perspective of CALL and in the light of RAPIDO's fieldwork in Europe. It proposes a relatively simple and transparent evaluation system which takes account of both the functional characteristics of ATs and the pragmatic features of their use.

2. Authoring Tools

The essential purpose of what we shall place under the generic label of Authoring Tools is the facilitation of software production. As such, ATs fit into a large class of products ranging from high level programming languages (such as *Visual Basic* which are facilitative in so much as they are easier to use than low-level languages) through so-called educational authoring "languages" like *Icon Author* and *ToolBook* (in which a lot of recurrent routines, like matching student responses or keeping track of their results, are pre-programmed but require to be linked by quite sophisticated scripting) to authoring "systems" ranging from the relatively complex *Authorware*, to *Speaker* and *Partner Tools*, which require no programming at all; these rely upon menu-driven selections by the author, or the use of templates, with text or media objects being slotted in appropriately.

An important reason for using the generic term Authoring Tool is that it encourages the expectation that individual tools are part of a larger tool-set at the disposal of the educational handyman. It takes us away from the dated notion that a single "system" is the answer to all courseware contingencies. It helps prepare us for the flexibility that intranet-delivered courseware will demand of its authors, where each element in a course may well be driven by a separately constructed program using appropriate tools.

When RAPIDO surveyed the product range in 1996 it initially found 259 self proclaimed ATs on the world market which might have relevance to CALL. 80% of these were eliminated using simple criteria:

- running under Windows 3.11 or above on the PC platform and/or under OS 7.52 on the Apple Macintosh platform
- incorporating audio and/or video capability
- possessing a real presence in the European educational market.

Since 1996, WWW-delivered ATs have extended the range of products available to CALL developers, and, as a result, the market segmentation between Europe and the US is becoming blurred. Information on 29 products has been collated for the present paper.[3] These include one important new category of AT: tools which allow Web-based courses to be created and delivered over the net, ranging from exercise generators to what have been termed course or classroom management systems (Graziadei et al. 1996). They have the distinct advantage of being platform independent.

3. Approaches to evaluation

This chapter will limit itself to the complex yet fundamental issue of the basis upon which a CALL author may select tools from such a product range. We shall not attempt here to report in full on that product range, although examples will be given. Rather we shall consider previous approaches to AT evaluation and describe the solutions reached by RAPIDO and encapsulated in the Evaluation Criteria which were first applied to some key products in 1997.

One consequence of the growth in functionality of computing hardware and networks, and of the authoring tools designed to exploit this functionality, is that evaluation has become more complex. Currently there are two main approaches to evaluation: taxonomic and implementational.

3.1 Taxonomic evaluation

Taxonomic evaluation establishes lists of possible or desirable features, and checks whether tools possess them. The result provides a score, and this converts into a rating. The lists can be derived either from manufacturers' product specifications of what tools are supposed to be able to do (like "provide password security", or "present information in the form of on-line glossary support" (Landon 1999)) or they may be based upon an analysis of what tools are typically used for (working from the features implemented in real applications). Or, finally, a list can be derived from what an author might be observed doing, such as "changing text spacing", or "effecting audio channel control". There is clearly a fair degree of overlap between these taxonomies, and they have been criticised for arbitrariness, narrowness and circularity.[4] Locatis et al. (1991) rightly point out that they do not evaluate the effort involved in combining a multiplicity of features to form a real application.

Taxonomic evaluations can be summarised (drawing upon Locatis et al. 1991, and MacKnight & Balagopalan 1989) as falling into four main categories:

- *Feature Checklists* (using a simple, and what can appear an arbitrary taxonomy of desirable features derived from an established notion of what constitutes an ideal authoring tool)
- *Weighted Feature Checklists* (which is the same, but has the sophistication of weighting the features according to a value system)
- *Lesson Element Checklist* (identifying the key features found in (parts of) lessons, estimating their frequency in any normal application, and measuring

the keystrokes, time and mental effort to author them)
- *Authoring Task Inventory* (employing what seeks to be an exhaustive taxonomy of authoring tasks that can occur in any broad functional area such as
  "handling text objects").

Different authors have identified quite different sets of functional area. For Yeager
(1994) these consist of four broad categories:

1   presenting information (text, graphics, audio, video)
2   asking questions/judging answers
3   creating branches, both navigational (based on student choice) and condi-
    tional (based on student response)
4   saving data (for evaluation purposes).

MacKnight & Balagopalan (1989) identify at least ten broad functional areas or
commonly used instructional components: authoring environment, text creation,
text editing, graphics creation, graphics editing, animation, interactive video, instructional strategy, calculations, and student management. They add that interactive sound and artificial intelligence constitute new but as yet unmapped functional
areas (op. cit.: 1232). Bruce Landon, evaluating on-line delivery software for the
Canadian user, has a similar range but rightly extends it to "potential for collaboration and connectivity" (Landon 1999). NSTL (Software Digest/PC Digest) have a
concept of "Versatility", by which we understand "range"; this comprises General
Features, Cross-Platform Compatibility, Applications Development, Text Tools,
Drawing and Image Tools, Animation and Video Tools, Audio Tools, Internet Tools,
and Miscellaneous.
    A list of 6 broad functional areas was applied by the RAPIDO Project, namely:

- *Macro Features* (hardware, networking and broad difficulty rating)
- *General Features* (system, authoring environment, programming facilities,
  author ergonomy, learner ergonomy)
- *Handling Objects* (at macro level (including hypertexts), text strings, graph-
  ics, animation, audio, video)
- *Learning and Teaching Functions* (learner inputs, feedback, navigation, test-
  ing)
- *Management* (of student, of group, performance data, security)
- *Product Ecology* (documentation, training, support, user groups, licensing,
  hardware flexibility, product strategy, credibility).

This is effectively an expanded Authoring Task Inventory, further extended, where
appropriate, to evaluate learner tasks. It creates 211 fields of data for author functions and 106 for learner functions, making a maximum score[5] for any single Authoring Tool of 634 points (0, 1 or 2 points are allocated per field in the current
version).

It has been pointed out that broad functional areas, let alone specific task functions, cannot stand alone but must coalesce in an application in such a way as to display what may be termed a macrofeature such as "clarity". Thus, Gupta & Haga (1994) stress the importance of integration, Park & Seidel (1989) look for good interfacing (which goes hand in hand with good integration), and Filipczak (1995) emphasises communication. Finally, it has also been suggested by Park & Seidel that the most important authoring tasks performed by authors are design-related, calling for a qualitative judgement rather than a taxonomic quantification.

Such caveats can be accommodated in part by devising appropriate weighting systems for feature checklists, but it must be accepted that a taxonomy can never escape the imprint of its own didactic philosophy: it is always open to rebuttal on what might be called meta-methodological grounds. It is also necessary, if macrofeatures like "integration" and "communication" and "interface" are to be evaluated, that they be defined in terms of observable and discrete phenomena, and this appears difficult to accomplish.

3.2 Implementational evaluation
Given the shortcomings of a taxonomic approach, evaluators have commonly experimented with what may be termed implementational evaluation involving various forms of lesson or task benchmarking. Sample lessons or applications are specified, and then executed using different tools, with a measure being taken of the time and cost of each implementation (e.g. Riley 1995 and Bickerton et al. 1997). This approach is widely used, especially in commercial evaluations, but it suffers from a number of drawbacks. It is not easy to specify which tasks can, realistically, be performed by all the tools under investigation. A common denominator task is non-discriminatory, whereas the non-common capacity of an AT may be precisely what an author needs. And what happens if a tool cannot perform a task in part or in whole? Does one question the specification of the task, or try and evaluate the imperfection, and if so, how? It has been pointed out (Locatis et al. 1991) that the results of lesson bench-marking are particularly user-dependent, especially for ATs which involve programming or scripting. This is seen clearly in a variant on the above implementational approach which involves the observation of the authoring process in real time, with recordings being made of authors' outputs (O'Malley et al. 1991).

Neither of the above main approaches to AT evaluation is without weaknesses. A programmatic view of the learning process will tend to stress structured features at the expense of freedom to navigate; a grammar/translation approach to language learning could lead to tasks which favour text manipulation at the expense of, say, media object manipulation; and a constructivist method will evaluate more highly those features which allow students to explore and analyse data than activities requiring "correct" responses. None allows one to judge what Wulfekule (1994) claims to be crucial, namely the adaptability of the application to the learner.[6] The RAPIDO approach to evaluation sought to combine an extensive taxonomic scheme (covering both author and learner functions) with a fairly simple lesson benchmarking process. The results are reviewed below. But first we must consider additional contextual parameters in the authoring process which have fundamental implications for the evaluation of ATs.

4. Pragmatic parameters in evaluation

4.1 Use

A recognised feature of language education is its methodological diversity and hence its wide range of purposes. Teachers and learners can have quite different aims and contexts, and CALL applications need not be powerful to be effective. Simple may be beautiful, powerful may be pointless[7]. Different use-related taxonomies have been suggested. Some focus on instructional design[8], whereas others use the delivery system (platform, media facilities, Web or connectivity) as the main constraint upon use, and hence a substitute for a typology of use. We have opted for the following scheme that reflects an AT's degree of generality or specialism but that is also hospitable to Web-based delivery systems.[9]

Generic MM Authoring Tools
1   Multimedia Application Developer
2   Educational Courseware Developer
3a   Course/Class Management Tools
3b   Ditto but Web-delivered

Generic CALL Authoring Tools
4   Stand-alone CALL Developer
5   Language Laboratory Emulators
6   Language Course Management Tools

Specialised Authoring Tools
7a   Test Developer
7b   Ditto but Web-delivered
8   Hypertext Developer
9   Presentation Designers

It is also customary to evaluate products in terms of a fitness-for-purpose criterion which goes beyond the above categorisation, and clearly the only sure way of evaluating ATs in terms of such an open-ended concept would be to characterise all possible classes of use and relate these to a features checklist. In practice, this is an excessively complex and cumbersome procedure, since there can be alternative ways of achieving a declared purpose, particularly with the use of some ingenuity. It is more simple, therefore, to rely upon two other techniques.

First, as we shall see, there is a loose correlation between range and fitness for purpose. The more features an AT possesses, the wider its potential applications.

The fact, for example, that product X has a feature for "showing or hiding subtitles, under author/learner control, over a video object", may be a feature not possessed by other tools and one which allows greater scope in developing learning activities. This feature enhances the range of tool X, its use-related features are consequently more extensive, and the tool may be said to be potentially more effective.

Second, a judicious use of task benchmarking can reveal how ATs perform against a range of characteristic tasks. This technique does have limitations, since users vary in their ingenuity and tolerance, and tasks vary in their clarity of definition. Riley created a task benchmark based upon an implementation of "the same semi-functional multi-window interface" for each of the ATs selected for review. The screen was divided into windows or panels designed to display video (in .avi format), image, text, controls (icons representing buttons for navigation and for audio (.wav) replay), and a context sensitive instruction panel (Riley 1995:3).

The principal drawback of such a benchmarking system is that it is divorced from the requirements of the language teacher. For this reason RAPIDO established an inventory of five tests, each with a significant language learning focus:

1   Import text and generate a hypertextual link between it and a series of objects (text, picture, sound, animation and a video sequence).
2   Extract 4 stills from a video sequence, randomise them and permit a user to re-order them.
3   Import a sound file for each of the above stills, attach it to each of the pictures created in Test 2, and permit a user to retrieve any of the sound files.
4   Emulate a user by moving a sound bite into two structured storage environments (e.g. notebook or database).
5   Add sub-titles to a 20-second sound or video sequence.

Even here, it proved difficult to produce genuinely comparable outcomes, and it would be unreasonable to base the selection of an AT uniquely upon the variable quality of evaluators' results. One may conclude, however, that it is important to place value upon task benchmarking results only in so far as these do relate to an appropriate set of pedagogical objectives. And any such judgement of an AT must also be qualified by reference to its global feature range, as established by traditional taxonomic evaluation.

4.2 Users

In the world of CALL, potential authors vary from linguaphobic technologists to technophobic linguists, from polyglot programmers to technomanic teachers. The degree of facilitation offered by a given authoring tool – and this purpose is the one common feature of all tools – will evidently vary according to the skills and environment of the user. Vendors understandably emphasise their products' range or power, ease of use ("requires no programming", "no use of CGIs or HTML"), enhanced development speed and the resulting cost reductions or efficiency gains. It is often implied that such benefits characterise the Authoring Tools themselves ("this product is a joy to use") whereas in fact only range is a function of the tool itself; all other features are user-determined, and users, like learners, are all different.

Three distinct categories of user were identified in the RAPIDO analysis, those able to use difficult, medium or easy ATs. Difficult tools were defined as requiring significant amounts of programming expertise; medium tools required scripting abilities which go beyond the manipulation of templates, and a level of complexity (often occasioned by the range of features supported) which takes the AT beyond the bounds of non-technical use; easy tools required no programming, scripting, or significant manipulation.

Evaluators reported no great difficulty in applying these categories. It is arguable, however, that a subdivision is needed for the "easy" ATs, and this observation is based upon further trial and error. There are those easy ATs which require a fair degree of technical ability to set up and operate, possibly through poor or complicated design or hardware contingencies, and those which live up to their claims to be "plug and play".

On this basis the products currently listed by the RAPIDO Project[10] fall into the following user bands. The use-related codings which are given here are those proposed in section 4.1; they range from powerful generic development tools (coded 1), through Language Laboratory emulations (5) to specialised application tools (9):

Easy: plug and play

Acrobat (Adobe)	(9)
GapKit 2.0 (Camsoft)	(7a)
PowerPoint (Microsoft)	(9)
Question Mark 3 (Question Mark Computing Ltd.)	(7)
Translt-TIGER Authoring Shell (TELL Consortium and Hodder & Stoughton)	(8)
Wida Authoring Suite (Wida Software Ltd.)	(4)

Easy: with technical support

Apple Multimedia Learning System (Apple Computer Inc.)	(5, 9)[11]
CAN-8, with EAASYII (Sounds Virtual)	(5)
Ematech 3.1 (Acti-Média systems)	(7a, 9)
First Class (SoftArc Inc.)	(3b)
Hyperlab 2.5.1 (Multimedia Diffusion)	(5, 6)
Learning Space 1.1 (Learning Labs)	(6)
Partner Tools (Teleste Educational)	(4, 6, 7a)
QuarkImmedia (Quark Systems Ltd)	(9)
Speaker Author / Speaker Creator (Neuro Diffusion)	(4, 6)
Symposium (Centra)	(3b)
Top Class (WBT Systems)	(3b)

Medium

Authorware 4.0.1 (Macromedia)	(2)
Guide (Owl International)	(8)
Hypercard (Apple Macintosh (also PC version)	(8)
HyperStudio (Roger Wagner Publishing)	(8)
IconAuthor 7.5 (AimTech/Asymetrix)	(2)
LAVAC 6.02B (C Puissance 3 Informatique)	(4, 5, 6)
Quest 5.1 (Allen Communication)	(1)
Supercard (Silicon Beach Software)	(8)
ToolBook II Instructor/Assistant (Asymetrix Corporation)	(2)
WinCalis (Duke University)	(4)

Difficult

Director 6 (e.g. + Adobe Première) (Macromedia)	(1)

5. Quantifying evaluation

The final stage in an evaluation is to derive a means of expressing simply how different characteristics of Authoring Tools perform. Taking range, ease of use and cost as the prime values,[17] accompanied by the data from the use and user dimensions identified above, a combination of taxonomic or implementational measurements can be set in place together with the qualitative judgements of use and user. Such an approach was partly adopted in principle in the RAPIDO Project, but its application to more than a few key ATs has been limited by funding constraints.

5.1 Range
The functional range of a tool can be determined by rating it against a checklist. Tools with the widest repertoire of features possess the greatest range. In the RAPIDO scheme features are counted both for author and learner, producing a total of 317 points for adequate and 634 for good performance. The results for five leading authoring tools with a significant user base in language education are given below in Table 1.

An alternative way of expressing range is in terms of the distinction noted above between the following types of AT: generic courseware development systems (8 in number); generic CALL ATs, including language laboratory emulators (10), and specialised ATs (11). Clearly, one would expect that the more specialised the category, the lower the score.

5.2 Ease of use
To calculate ease of use, it proved appropriate to combine two measurements. Evaluators using the taxonomic checklist were asked to give a global "difficulty" rating for the tool, using the three categories (easy, medium and difficult) described in section 4.2 above. Secondly, evaluators were asked to exercise an additional quali-

tative judgement and grade a tool on any particular feature in terms of three scores for usability *by the user type* designated by the above difficulty rating:

0 = impossible/useless
1 = adequate/of some benefit
2 = works well/definitely worth using

As a consequence, "ease of use" appears both as a basis for establishing user-type categories ("suitable for users able to handle an easy AT", for example), and as a means of scoring the performance of features in the hands of designated types of users ("works well for users able to handle a medium AT", for example).

Table 1: Products by user type and performance

Product	User type	Score
Director	difficult	487
Authorware 3	medium	426
Speaker	easy	415
Lavac 4.01	easy	340
Partner Tools	easy	277

5.3 Cost

As for cost, the simplest approach is to take a measurement of the time needed for a suitably experienced author to execute the tasks as established in the implementational evaluations. It would appear that this result bears a close correlation to the keystroke count (which has a bearing on ease of use). However, additional cost features such as licence charges for the software, and hardware requirements, can be computed into the cost equation. Interestingly, O'Malley (1991) extended the evaluation process to the observation of authors in real time. This proved to be an effective but expensive and slow way of calculating ratings, and has not been widely used.

5.4 Use/User Dimension

As we have seen, any evaluation system that omits the pragmatic criteria of use and user will fall into the trap of simplistic globalisation which we criticised at the outset. The evaluation of an AT must therefore involve recognition of its insertion into a user/use matrix. This completes the data required by potential authors, for it complements a measure of the range of an AT with an evaluation of the author's own circumstances and objectives. Taking six products and applying a use/user matrix, we obtain (in Table 2 below) the kind of data we need if our selection of ATs for Multimedia CALL is to be judicious and effective.

It is no accident that ATs with a narrow feature range like *Partner Tools*, or a wider feature range like *Speaker*, having no programming or scripting features, score highly for ease of use. An AT like *LAVAC* which has the option of a limited programming language is somewhat harder to handle. Products like *Authorware*, *Toolbook* and *Director* require the user to employ programming systems, and are evaluated as being correspondingly harder to use. These products, in turn, are easier to use than the fully-fledged programming languages which, like *Visual Basic, C++* and *Java* fall outside the class of ATs as defined here.

Table 2: Products by use and user type (ranked by range score)

Type of use	Programmer	Scripter	Supported academic	Plug & play academic
(1) Generic MM Authoring	Director	-	-	-
(2) Educational Courseware	-	Authorware	-	-
(3) Generic CALL	-	-	Speaker Partner Tools	-
(5) Lang Lab Emulator	-	Lavac	-	-
(7a) Specialised Test Developer	-	-	-	GapKit

6. Conclusion

The range of tools on the market is large and constantly changing, and there is a sense in which the variety of pedagogic applications to which they can be put is limited more by the user's imagination than any intrinsic features. The selection of an AT can proceed pragmatically according to immediate availability, cheapness and least resistance, and users make a virtue of all sorts of necessity.

As soon as users plan to invest significant time and effort in authoring learning materials, however, an empirical analysis of "the best AT(s) for the job" seems desirable. We have seen that highly relevant data can quickly be collated using a feature list taxonomy such as developed by the RAPIDO team. It is clear that the feature range of an Authoring Tool is the indicator which is objectively the easiest to evaluate. It also has the advantage of correlating with intended use, since the greater the functional range of an AT the greater its potential sphere of application. By a simple measure of features one is therefore able to identify potential areas of application.

But there is also evidence that the greater the functional range of a product the greater its complexity and ultimately the need for the user to become engaged in programming operations. Qualitative data on the suitability of ATs for different user types, together with simple economic cost factors, can lay the basis for a more comprehensive evaluation. This may be as far as one can reasonably go before beginning to hit some of the imponderables in the AT / user / use equation. Are most features really necessary for successful pedagogy? Do the ATs support co-operative working between course designers and learners? What kind of academic linguists make good AT users? Do the products of authoring really justify the effort of production?

These and other issues must be addressed by applying different measures from those described here (Bickerton 2000), and they serve to highlight the essential complexity of the authoring act when performed by linguists for the benefit of language learners.

Notes

[1] *Rapid Authoring of Packages using Innovative Development tOols.* This SOCRATES Project (TM-LD-1995-1-GB-58 (Universities of Plymouth (co-ordinator), Antwerp and Grenoble, and Teleste Educational)), examined the feasibility of MMCALL authoring by 45 academic linguists supported by a team of trainers and technical advisors using field trials of a new set of ATs produced by Teleste. The Project reported in February 1997 (see Bickerton, Ginet, Stenton, Temmerman & Vasankari 1997). The Report is available from david.bickerton@pbs.plym.ac.uk.

[2] SMIL, Synchronised Multimedia Integration Language, the latest development of the W3C Synchronised Multimedia Working Group, promises to be an easy-to-learn tool for producing "cheap, effective and easily updated multimedia courseware, accessible worldwide. Users could view video, follow links, search topics and contact tutors" (*Times Higher Educational Supplement,* 14 November 1997).

[3] Only products which have reached the market place and established some kind of user base are listed. Products in development are omitted: e.g. *The Language Author* – a templating system for authoring dialogues, written using *AuthorWare* – (Bangs 1997). It must be recognised that many authoring tools are developed which never achieve significant market penetration and quickly become obsolete.

[4] They evaluate an AT on the basis of what ATs do rather than on an independent view of what they *should* do.

[5] I.e. for a product's performance (useless = 0, adequate = 1, or good = 2) judged against the feature checklist.

[6] The evaluation of learner-responsive or learner-adaptive features of ATs is absent in all taxonomies and implementations currently being used.

[7] G. Davies has reported that of all the ATs marketed in recent years, the simplest (Wida, as listed in note 10) have the greatest market penetration. This uptake pattern may confirm user rather than use-related characteristics, and it relies upon historical trends and an ageing technology.

[8] Bangs (1997:153–4) calls this "modes of interaction" and distinguishes presentation, frame-based (branching) and book-based (browsing) systems. These correspond roughly to types 9, 2 and 8 in our own typology.

[9] Web delivery is an alternative to the production of stand-alone or networked applications. We anticipate that whereas most products falling into category 5 today require hardware add-ons, they will soon be replaced by Web-delivered solutions.

[10] Not applicable (because written for DOS), although widely used, are the original ver-

sions of *Gapmaster, Choicemaster, Matchmaster, Storyboard, Testmaster, Pinpoint, Wordstore, Vocab* (all by Wida Software Ltd); also noted, but not widely known or available, are: *cT Professional* (WorldWired, Inc.), *DEMOquick Demo Generator* (AMT Learning Solutions, Inc.), *Interactive Assessor* (EQL International Ltd.), *Learn Link I-net* (ILINC), *Newspeak* (Ascola), *Pathway* (Solis), *PHOENIX for Windows* (Pathlore Software Corporation), *Rapid 3.5* (Emultek), *SMIL, Strata MediaForge.*

[11] AMLS is actually designed as a system for presenting video from multiple sources.
[12] MacKnight and Balagopalan (1989) use the terms "functionality", "flexibility" and "productivity" for these concepts.

References

Bangs, P. (1997). Author, author! – but does (s)he speak my language? In D. Little & Voss, B. (Eds), *Language centres: Planning for the new millennium, papers from the 4th CERCLES conference* (pp. 145–162). Plymouth: CercleS.

Bickerton, D., Ginet, A., Stenton, T., Temmerman, M. & Vasankari, T. (1997). *Final report of the RAPIDO project.* Plymouth, UK: University of Plymouth (SOCRATES Project TM-LD-1995-1-GB-58).

Bickerton, D. (1999). Authoring and the academic linguist: The challenge of MMCALL. In K. Cameron (Ed.), *CALL: Media, design & applications* (pp. 59–79). Lisse: Swets & Zeitlinger.

Bickerton, D. (2000). Can (and should) academic linguists become multimedia authors? In *Fremdsprachenlernen mit Multimedia [...], Triangle 17, 30–31 janvier 1998* (pp. 47–63). Paris: ENS Editions (for Goethe-Institut, ENS Fontenay/Saint-Cloud, The British Council).

Botto, F. et al. (1996). Reviews for Australian Consolidated Press. Available (2000), search: http://apcmag.com/reviews/default.htm

Brücher, K.H. (1993). On the performance and efficiency of authoring programs in CALL. *CALICO Journal, 11.2,* 5–20.

Filipczak, R. (1995). On the trail of better multimedia. *Training, 32.11,* 56–62.

Graziadei, W.D., Gallagher, S., Brown, R. & Sasiadek, J. (1996). Building asynchronous and synchronous teaching-learning environments: Exploring a course/classroom management system solution. Available (2000): http://horizon.unc.edu/projects/monograph/CD/Technological_Tools/Graziadei.asp

Gupta, U.G. & Haga, C. (1994). Evaluation of four PV-based authoring systems for business instruction. *The Journal of Computer Information Systems, 34.3,* 24–28.

Landon, B. (1999). Online educational delivery applications: A Web tool for comparative analysis. Available (2000): http://www.ctt.bc.ca/landonline

Locatis, C. Ullmer, E. & Carr, V. (1991). Authoring systems: An introduction and assessment. *Journal of Computing in Higher Education, 3.1,* 23–35.

MacKnight, C.B. & Balagopalan, S. (1989). An evaluation tool for measuring authoring system performance. *Communications of the ACM, 32.10,* 1231–1236.

O'Malley, C., Baker, M. & Elsom-Cook, M. (1991). The design and evaluation of a multimedia authoring system. *Computers and Education, 17.1,* 49–60.

Park, O-C. & Seidel, R. J. (1989). Evaluation criteria for selecting a CBI authoring system. *THE Journal (Technological Horizons in Education), 17.2,* 61–68.

Riley, F. (1995). Understanding IT: Developing multimedia courseware. Available (2000): http://www.hull.ac.uk/itti/hullprod.html

Strauss, R. (1995). The development tool fandango: Deciding the authoring system versus programming language question. *CD-ROM Professional, 8.2,* 47–55.

Yeager, R. (1994). Author! Author! *Training and Development, 48.5,* 97–104.

Wulfekule, N. (1994). Selecting a hypermedia authoring program for CBT. *THE Journal (Technological Horizons in Education), 21.7,* 77–80.

5

DISSEMINATE or not? Should we pursue a new direction: Looking for the "third way" in CALL development?

Philippe Delcloque
University of Abertay Dundee, Scotland, UK

1. Introduction

What follows is a presentation of a concept and architecture conceivably capable of becoming a development movement on a significant scale in CAL (computer-assisted learning), not just CALL (computer-assisted language learning). DISSEMINATE has been in gestation in the mind of the author for some three to four years and has now been presented in two official fora, the inaugural WorldCALL conference in Melbourne in 1998 and the EUROCALL 98 conference in Leuven.

This chapter will begin by describing the seeds of DISSEMINATE, the practical and theoretical background which came to influence the birth of the concept, especially its underpinnings in CALL practice in the last three decades. This will be followed by a possible theoretical justification for DISSEMINATE, a presentation of Gagnepain's theory or model of mediation and its possible relevance to CAL(L) development and delivery. In the third section the DISSEMINATE proposal will be explained through an "explosion" and explanation of its component parts. The conclusion will then focus on the reasons which might lead to the adoption of DISSEMINATE and its possible relationship with other "hypercollaborative" movements such as *GNU, Linux, WebCT*, but also various European initiatives as well as the UK-based TELL and WELL consortia. Work will then continue outside the scope of this first description to examine the initial specifications to DISSEMINATE, followed by the practical implementation of what by then may have become a project and/or movement.

2. The seeds of DISSEMINATE

The CALICO 97 conference had as its theme "Content, content, content". Was this not a veiled admission of what so many scholars have said for decades, namely that CALL had been influenced more by technology than pedagogy or content? As some practitioners in the field thought that the discipline was coming of age or at least reaching a degree of formality, the conference concluded with a debate on the value of authoring systems and a proposal for a monograph on the same topic. While this was happening, many were beginning to extol the virtues of the latest "bandwagon", the now maturing World Wide Web, which has brought yet another change of emphasis in CALL practices. The idea of the monograph died and Web-Enhanced Language Learning, however that may be defined, continued to expand on an exponential basis, which might suggest a cause and effect relationship with respect to the death of the monograph. Efforts in the field, it could be argued, had to be refocused and reviewed once again as a function of technology.

Increasingly, the necessity to have a deep reflection on the various components and actors in the process of CALL courseware production is particularly crucial as we reach a new turning point. This might involve an examination of content, media, enabling tools, physical (hardware) support, human factors (in the broadest possible sense), collaboration, implementation, evaluation and dissemination. The purpose of what follows is to continue to fuel this process of reflection and frame it within a particular collaborative proposal.

The famous quotation from Marshall McLuhan (1964)[1] "The medium is the message" has occasionally been described as a distortion of the words actually uttered by the pioneer of modern communication theory. It appears to be unclear whether he said: "The medium is the message", "The medium is a massage" or "The medium is the massage". In reality, these interpretations present slightly different viewpoints with the last two representing a greater distortion in the described process of content mediation. Levy, in his seminal work (1997a:2), points out that the distorting effect of technology in shaping final CALL courseware as well as influencing practice in development is extensively covered in the literature of the last 30 years with Kenning & Kenning (1990), Lian (1988) and Murison-Bowie (1993). The description given below is a more personal view of how CAL(L) practice was shaped along the historical axis as a series of phases. This diachronic description of the field must naturally be seen as a series of overlapping phases, rather than a series of movements, which stop and start as the next begins. Indeed some of the early phases still manifest themselves in late design.

Phase 1: The Pioneering Wilderness or Text Phase – 1960s and 1970s

Every field has its frontier, and the work during this period, especially in the early ten years had the features of the frontier of the Wild West: laborious, sometimes exciting and mostly dangerous. Extravagant claims were made about what computers could do in language learning – in time. At that particular point, computers were used to support lines of text in mostly man to machine communication and processing. Authors were invited to become programmers and to shape learning output within the severe constraints of the delivery medium.

Much of the pioneering work in CALL involved the manipulation of text on screen, largely driven by the inadequacies of the early widely available operating systems and/or programming languages: *BASIC* and Pre-*DOS* assembly languages. Some of that work remains amazingly relevant and useable despite its age, particularly in the field of practice drills for the written language (grammar). It should be remembered that the PLATO Project started in 1960, and *BASIC* was invented in 1963. As Levy (1997a:51) reminds us, much influence came from Skinnerism, which had underpinned the programmed learning approach, found in audio language laboratories. This later prompted the development of functional exercise development tools in the early 1980s, such as Wida's series of authoring packages[2] and Camsoft's *Fun with Texts*[3]. In emphasising the need for terminological consistency, essential at the inception of any science, some scholars may wish to adopt the expression "monomedium exerciseware" for the practice of this period. Crucially, at that point, CALL pioneers were creating enabling tools rather than courseware, believing that "ordinary" linguists would input their own materials. The actual content provided with the said tools was there purely to illustrate what could be authored with the tools.

Phase 2: The birth of a new platform, The Platform Phase – the 1980s
In 1984, with the advent of the Macintosh and a few years later, in 1987, with the launch of *Hypercard 1* bundled with every Mac, came the Apple & *Hypercard* phase. The importance of this was peripherally felt in the gaining market share of *Hypertalk* versus *BASIC* as a programming language in CALL development, and much more critically felt in the vital introduction of sound playback and recording into CALL programs. This created a new design generation for CALL work and a split along platform lines which, to some extent, is perpetuated to the present day with the difference between European development (largely PC-led) and North America and Australasia, (significantly Mac-led). The reason why this may be considered to have some theoretical significance is that it mirrors the early post-Second-World-War split in general and theoretical linguistics between the European tradition and the North American and Eastern traditions. That phase was also fundamentally important in bringing universal recognition to the supposedly new concept of "hypermedia", which led to the encouragement of student-driven and exploratory learning. The term "hypertext" had been coined first by Ted Nelson in 1965[4], and the Graphical User Interface (GUI), which was invented at Xerox Park, had preceded the advent of Apple computers.

Phase 3: The Laser (Videodisc) Phase
This was mostly a North American phenomenon, which harnessed well the power of the television screen picture ("mediatainment") to produce more lifelike immersion. It was particularly exploited for the purpose of audiovisual comprehension rather than for interaction through expression or production, given the continuing

computational limitations of learner feedback quality. One of the best examples was the *A la rencontre de Philippe* adventure courseware, which formed part of the valuable Athena Project at the Massachusetts Institute of Technology (MIT) led by Janet Murray (1995), the narrative specialist, and AI expert Sue Felshin (1995). Other examples include *Montevidisco* (Schneider & Bennion 1984) and *Expodisc* (Davies 1989; Davies 1991).

Phase 4: The Multiple- Multimedia Phase

Technology-driven (rather than enhanced) CALL continued in the views of many with a near obsession with producing glitzy multimedia. The quality of linguistic content appeared somewhat sidelined in favour of the inclusion of the most up-to-date media assets such as moving pictures, fashionable animations, video production and semi-broadcasting techniques. There appeared to be little attention or reflection paid to the role and function of each component media and to the overall pedagogic design. Some scholars began to use the term "multiple media" rather than "multimedia", although it is difficult to establish who might have been responsible for coining the expression "multiple media". The distinction is important because many linguists involved in CALL felt that an appropriate use of multiple media in autonomous learning situations might be the answer to combating constraints of time and space and accommodating learning styles in the context of total or semi-autonomous learning situations.

Phase 5: The CD-ROM Phase

The next evolution (rather than revolution) came from storage (and to some extent associated RAM) developments. A whole new generation of complete courses on main platforms designed for use in self-access and/or as classroom teacher-led tools appeared. This was a departure from the more commonly available functional exercise tools described above ("monomedium exerciseware"), as the market was now dominated by so-called complete and comprehensive courseware solutions designed for a particular language level (beginners, intermediate, etc.). The same split as in conventional text-book production was found in the "crude distinction" between the tenets of total immersion (recourse only to the target language) and the contrastivists (who worked with source and target language pairs). One major design failure of this period was its "commercial neglect" of authentic cultural immersion issues. A language CD-ROM would be distributed for the teaching/learning of four different European or Eastern languages using the same databank of pictures drawn from North American digital libraries and the same pedagogical paradigms.

During Phase 5, much attention had been given to the use of authoring tools used in CAL(L) and multimedia production. These included *AuthorWare*, *Guide*, *Hypercard*, *Toolbook*, *Visual Basic* and many others. Once again the CALL enthusiastic linguist was asked to accept a somewhat lengthy learning curve to create materials for bespoke or commercially related purposes.

Phase 6: The dedicated language teaching and learning Authoring Tool Phase

The natural progression from Phase 5 was to construct authoring tools that were dedicated to producing language learning and teaching courseware. In some ways, this was a return to the early "exerciseware" efforts except that these were fully integrated suites based on a number of paradigmatic templates. The term "paradigm" as against "toolbox" has been coined by Hart (1981:7) in his description of the PLATO project.[5] Toolbox tools are open visual coding devices such as those mentioned in the preceding section. They allow authors to freely design their screen and involve a longer learning curve. Paradigmatic tools are based on fixed templates that allow rapid authoring and a lesser learning curve but lack flexibility. Examples of such tools include *Lavac*, *Libra*, *MacLang*, *SuperMacLang*, *Speaker* and *WinCalis* to name just a very few. Wida's series of authoring packages (see above) have also been given a front end and turned into a comprehensive authoring tool known as *The Authoring Suite* (see Endnote 2), so that they can now be included in that category. Information on the effectiveness of Phase 5 and 6 tools can be found as part of the data in the RAPIDO report published by Bickerton et al. (1997) and in the chapter by Bickerton et al. in this publication. See also the publications quoted in Levy (1997a:257) and in Levy's CALL *Annotated Bibliography* (1997b).

The advent of Bertin's *Learning Labs*[6] and our own somewhat misguided effort in the *Authorlang Prototype*[7] came at a time when, according to Graham Davies[8], a situation of over-supply of such tools was meeting an ever-decreasing demand. *Learning Labs* was originally known as *Learning Space*, a name that had already been adopted for the large-scale courseware tool development project undertaken by IBM/Lotus[9,] which parallels the valuable co-operative *WebCT*[10] development. The virtue of the said tools however was to cement the useful dimension of comprehensive learner monitoring/tracking through networks which can be used for research evaluation purposes and for adaptive techniques.

Phase 7: The Speech Intelligence Phase

As progress was made in the area of speaker-independent speech recognition and as linguists and engineers realised that many authentic linguistic situations involved closed predictable interactions, it became possible to harness the technology to introduce a new element to the person-machine interaction: speech recognition facilitating logically conditional interactions (branching dialogues), which produced a semblance of conversational realism and engendered a new type of drill capable of improving spoken comprehension and expression skills, a dimension often neglected by CALL work hitherto and now outperforming the traditional audio laboratory. In the eyes of some at least, "intelligent" CALL (ICALL) was beginning to see the light of day! The widespread commercial success of systems produced by Auralog[11] and Syracuse[12], to name just two major world players, bears witness to the growing acceptance of such advanced technologies in the CALL interface.

Phase 8: The Natural Language Processing Phase
This is perhaps the least successful stage, to some extent the phase which never truly happened, despite potentially offering the most promise in the long term. Some of the best attempts at harnessing NLP techniques were carried out at MIT under Athena. A good summary of what has happened up to the late 1980s in ICALL is found in Matthews's bibliography (Matthews 1991). The crucial point is whether intelligent analysis of students' responses can be made in real time in sufficiently free or open contexts. Guidance and inspiration from areas such as machine-assisted translation, translation memory systems and terminological databanks, would, in my view, provide a useful starting point rather than relying entirely on machine processing and generation (AI and expert systems). It could be said, however, that the practical outcomes of this phase are yet to come.

Conclusion to the diachronic picture
Throughout these formative years, few developers created teaching techniques drawn from the new media, or combined with increased effectiveness the advantages of various aspects of the technology. The prevalent approach appeared to be to transfer content from another medium (such as a textbook or audiotape) to the monomedium or multimedia computer. This extended through to exercise types (Cloze, MCQ, etc.). Some courseware tried to break new ground by encouraging exploratory learning and exploiting some of the areas which had dominated the progression of technology (games, especially of the adventure type, simulations, concordances) and other techniques which might be said to truly belong in the digital domain.

What has become abundantly clear is that there were too few occasions when content drove the design and tuition process. The drive came instead from the general progression of associated technologies, platforms, tools and other media. McLuhan also said in a television interview: "I am resolutely opposed to all innovation, all change. But I am determined to understand what is happening, because I don't choose just to sit and let the juggernaut roll over me." Perhaps the juggernaut of technology has obscured the obvious fact that first and foremost we are and will remain language pedagogues in a quest for increased effectiveness. As Levy (1997a:232) puts it, for the "average linguist", involvement in CALL development must be a matter of "cost versus gain", both in personal terms and in terms of collective added value.

3. Gagnepain's theory of mediation: A possible theoretical justification for DISSEMINATE

The epistemological theory of a little known French linguist may help us to understand the levels of mediation which pertain both in the authoring process and in the process of delivery of courseware to the end user.

On the basis of the clinical observation of neurological and psychiatric behaviour disorders over a period spanning more than 30 years, Jean Gagnepain[13] with the help of Olivier Sabouraud, a neurologist, posits four levels of mediation of human organisation, abstraction and behaviour. At the first (linguistic) level, the unit of analysis is the *sign* with its *langue/parole* or competence/performance characteristics, the sign being the underlying abstraction, as in De Saussure, and the symbol its designation (what it refers to in the real world). At the time when this research began, disorders such as aphasia were described largely in terms of the dichotomy between the Wernicke type and the Broca type with paradigmatic and syntagmatic realisations. The empirical research carried out by Gagnepain and Sabouraud's team, summarised in his seminal work in three volumes, appears to demonstrate that speech disorders can be categorised much more subtly between aphasia (disorder of the sign) and schizaphasia (disorder of the designation/symbol). Moreover both types show phonological and semiological dimensions (the term used by Gagnepain to refer to higher linguistic levels than phonology), and aphasia can indeed be split between the Wernicke and Broca varieties, confirmation of the existence of the two linguistic axes of selection and combination.

One of the immediate insights of Gagnepain's theory was the hypothesised existence of the second plane of "technological mediation" and the fact that dyslexia might indeed be not fundamentally a linguistic disorder but a disfunctioning of the manipulation and codification of skills and techniques or tools, technology in its original sense (Gagnepain 1982). On this second, technological, level, the unit of analysis is the *tool*, with disorders resulting in the absence of the underlying framework of representation/codification, atechnia (of which dyslexia might be the traditional linguistic manifestation), and disorders in the reinvestment of this underlying system in the real world (schizotechnia). This may have important consequences for the acquisition of technology in the modern world. It might explain why there are significant types, in particular the techno-anxious, or computer-anxious, and might account, reinforced through the conditioning of early education and traditional differentiation between sexes, for the relatively small uptake of women in ICT and the higher level of success, irrespective of gender, of the more transparent and intuitive, less codified, interfaces. Is it not strange that women, who have been shown genetically to be more natural and effective communicators, should still represent a very small proportion of Internet users? The Internet is about communication but it is mediated through technology. Relatively few women pursue careers as engineers.

The third and fourth levels pertain to the psychiatric domains, with the *person* as the unit of analysis on the sociological level, with resulting disorders such as perversions, psychoses, schizophrenia and paranoia and the *norm* as the unit of analysis on the ethical level where we encounter neuroses and psychopathies such as obsession and hysteria.

What is important within this framework is that when we use technology to teach/learn languages, we introduce at least two levels of human behaviour and knowledge/skill acquisition, the linguistic and the technological, and when we produce CALL Courseware, we may well be interacting at all four levels, given group dynamics and cultural issues with a small and/or capital C.

Gagnepain's theory has parallels in the machine interface and courseware engineering, which can be said to involve a superimposition of mediating levels:

- The hardware (this may be complex as it may involve more than one level such as the stand alone computer and a separate drive, such as CD-ROM, Zip or network)
- The operating system
- The authoring tool or code or Web browser (this sometimes means more than one level)
- The exercise paradigm or template (optional).

In addition, more levels mediate the end-product delivered to the consumer:

- The medium of delivery: multiple media package (textbook, audio, video, computer disc – including CD-ROM, monomedium solution, local or distant network, etc.)
- The author(s)' pedagogic design
- Peer interaction and generally what other writers describe under the label "social computing".

The effect of these various levels of mediation is to create "digital signatures" in the form of a recognisable operating system and/or authoring tool and/or particular human author(s) being perceived in the look and feel of a particular piece of courseware. It is essential to remember that, even in a learner-centred and resource-based environment, learning is going to be largely driven and mediated by the human tutor and monitor. Courseware is never neutral. It can be described as a multi-level mediated unit of "knowledge or information".

From this, we must be able to reach a number of methodological conclusions. Firstly, many feel that the learning interface must become as transparent as possible, so that the linguistic transcends the technological, the message transcends the media, the content transcends the support. Secondly, we might want to make our programs as "authentic and lifelike" as possible in terms of the nature of the interaction which they encourage. We might want to discourage the notion of excessively mediated interaction and use the power and progress of intelligent technology to create boundaries which are felt far less between man and machine. This, in turn, implies that we must work across disciplines and across neo-linguistic domains. This has happened on a number of occasions, with perhaps the best example found at MIT in the mid-1980s. Finally, we must share and build on our knowledge and work collaboratively to facilitate the ownership of tools with built-in flexibility instead of inbuilt obsolescence. To achieve this, the authoring tools we use and the courseware we develop must be authorable and expandable, going far beyond the concepts of open architecture à la *Opendoc* or *OLE*, both of which are proprietary.

Our grand scheme might be to empower willing authors with a shareable, robust, evolutionary, accessible mechanism for effective CAL(L) production and design, so that content creators may at last lead with the assistance of technologists. Examples of such development have occurred at the operating system level with the *GNU* and *Linux* developments.[14]

It is also the constraining effect of this mediation which must be examined. Much of the effort made in the last three decades has been guided by lowest common denominators (real or perceived) in the marketplace. This might have been DOS, low RAM machines and many other factors. The new risk is that development could be constrained to the perceived capabilities of Web authoring and delivery, especially on a distant network basis (Internet rather than Intranet or Extranet).

DISSEMINATE may provide some answers to combining some of the lessons we have learnt from past developments, achievements and failures.

4. DISSEMINATE explained

DISSEMINATE is a mnemonic acronymic vehicle capable of brand recognition to support a set of development principles for what some have referred to as the emerging science of Courseware Engineering (Marshall 1997). At a time when a major initiative in Europe is being proposed and a number of funded projects are heading in somewhat similar directions, it may be considered timely.[15] Above all, DISSEMINATE is an active verb which implies that the proponents may wish to follow this with definite planned actions which may occur simultaneously and in a coordinated fashion in all parts of the world. Finally, the exploded parts of DISSEMINATE about to be explained matter less than the sum and espouse one of the key elements of the concept: authorability, thus the definition of each component as proposed below is initial and flexible. It is not meant to be prescriptive and to create new constraints. It may lead to substitutions and refinements to suit individual authors and collaborators.

- *D* stands for *Distributed*, so that the fruits and tools of the project/movement are widely available and accessible across disciplines, nations, platforms and sectors.
- *I* stands for *Integrated* so that the component parts do not act as non-compatible entities, but indeed integrate with both existing and future technology and pedagogy. This also applies to possible intelligent functionality enhancements.
- *S* stands for *Stable* or robust and durable, perhaps the hardest attribute in the context of a development system with seriously dispersed ownership, but once again we may be guided by the relative success of Linux which now boasts ten million users worldwide across three operating systems.
- *S* for *Superimposed*, in the sense that the development is highly likely to be placed above the level of the Web browser without suffering from its limita-

tions. This might be labelled the top down version of DISSEMINATE (from the Web down), whereas the original intention was bottom up through the construction of a "vertical ladder" which would take authors from conventional authoring systems used to produce DISSEMINATE modules with full Web-compatibility. This is indeed the path which many major producers of authoring programs have chosen, including *AuthorWare, Director* and *Question Mark.*

- *E* for *Evolutionary*, perhaps one of the most important attributes to avoid the pitfalls of development which dies when a particular operating system or authoring program falls out of favour or becomes technologically obsolete. This is encouraged by open coding.
- *M* for *Modulaware*, responding to the evolution of modern computing away from monolithic multifunctional large entities towards ultimate discrete object orientation where the primary unit of construction or building block might be the Web screen or exerciseware module. In either context, data and assets would be entered easily and separately whilst the frame would contain all functionalities relying on core Web technology (*Java* applets, *Perl* code/scripts, *Javascripts* and CGI forms, functionality plug-ins including *Active X*) all controlled within non compiled authorable HTML and platform-independent virtual byte-code. The importance of the module is that it reintroduces the useful timesaving paradigmatic dimension, and stands alone for small developments.
- *I* for *Interactive* may be seen as more than a truism in the sense that much existing Web-enhanced learning might be said to encourage more passive acquisition given the perceived technical limitations of the medium. As interactive learning has now been the norm in many classrooms for several decades, it is important not to lose the foundations established through the best disc-based courseware. In the current state of the art, conventional CALL courseware is a great deal more interactive than Internet-based CALL courseware.
- *N* for *Networked*: In CAL(L), the last years of this millennium and developments into the third millennium might be said to have placed emphasis on the prominence of networks rather than stand-alone formats. This is mirrored in the real world where the delivery of products increasingly involves such structures which allow just in time distant access.
- *A* for *Authorable*: Always core within this concept and development architecture, this feature is one of the key elements, although not new, in the sense that it allows flexible rapid customisation as well as the much needed adaptation and instant error corrections sadly absent in fixed and closed systems. This is a movement away from obsolete staticity, inflexibility and proprietary tendencies.
- *T* for *Tracking*: Although tracking learners is rarely done for a whole variety of reasons (time, ethics, etc.), it is the necessary condition behind good monitoring and communication with the end-user, as well as the essential research instrument behind quantitative and qualitative evaluation. In the current state of knowledge, tracking may require the use of separate proprietary software,

although it is implemented in systems such as WebCT which will serve as a model and possible framework within which to situate DISSEMINATE.

- *E* must obviously end the acronym with the word *Education* although some might prefer the more fashionable *Edutainment*. There is certainly a well-documented need for the fun element in courseware, which has mostly been implemented by commercial companies culminating recently in the much praised *Oscar Lake* series.[16]

5. Conclusion: Why DISSEMINATE and how?

The architecture and potential movement might lead CAL(L) practitioners to build on past achievements (*Construction sur les acquis*) and continue into the future on a fully evolutionary basis. It might avoid a repetition of the mistakes of the past and provide a framework for ongoing continuous improvements. The advantages of the proposal might include:

- Tutor empowerment: intuitivity, rapidity, fast return on time and financial investment. High gain versus cost ratio. Possible genuine economies of scale.
- Facilitation of intelligent enhancements from Natural Language Processing both on the speech and text side.
- In time, a better fit between CALL and the functional, transactional, communicative approach. This is perhaps the most important question of all, usually discussed by talking about the place of CALL (classroom, self-access/study, learner-centred, tutor-centred, location-centred, process-centred [actions], peer centred [collaborative]).
- A higher concentration on the specificity of the interface? Exercises steeped in the digital domain: games, adventures, simulations, enhanced reality, object movements, conditional interactions, etc.
- Encouragement to creativity, the theme of the first WorldCALL conference and arguably one of the most important elements in quality authoring. Parallel examples exist in the music business where the freeing of the creativity of the masses through a movement away from virtuosity towards ease of programming (music sequencing) put composition within the reach of a far greater number of potential authors and creators.
- Avoiding the reinvention of the wheel: it is highly probable that Web-based language courseware will now involve much time-consuming duplication, with practitioners producing similar basic functionality exerciseware screens on an ad-hoc basis and repeating poorly what has already been done well in conventional software engineering.
- Due to open-coding and possible customisation, the need for strict specifications and standards becomes far less acute than in closed proprietary systems.
- The proposal would be usefully framed within a European or better world standard movement. It certainly would be framed within any widely accepted

Internet for Learning standard. In this context, it is worth noting that Java is being submitted to the International Standards committee.

- Similarly, open architecture implies compatibility with any theory-inspired instructional or pedagogic design. Given that there are at least 40 different approaches to second language acquisition, it is very clear that no model will ever be accepted as the dominant design paradigm.
- The system provides a useful "vertical ladder" between conventional and Internet browser-based authoring.
- It also provides a natural bridge between cognate disciplines within the field of language studies and scholars working on main platforms (Unix, Macs and IBM PCs). This could encourage the much-needed cross-fertilisation which has hitherto largely eluded CALL.
- The cross-fertilisation bridge may also operate across disciplines involved in CAL rather than CALL and encourage multidisciplinary teams to produce courseware for languages for specific purposes.
- Due to its extremely low cost in any event (the tools may even be totally free), the system is likely to be highly accessible and reach critical mass.
- It may allow CAL(L) practitioners to free more time to do research into more credit-worthy areas of CAL(L).
- It provides an answer to the "not invented here syndrome" because it allows the end-tutor user to modify existing materials to suit his/her specific teaching situation.

The disadvantages of the proposal have for me faded as the concept has matured. I used to feel that the biggest blocks to its acceptance would be its utopian nature and initially provocatively labelled it the CAL Esperanto. Some of the anticipated objections, which were never in fact formulated, might have included the fact that practitioners would not agree on the framework and its specifications, that substantial funds would be needed to accomplish it, that the system was bound to be too anarchic and unstable, that issues relating to copyright and intellectual property rights generally could not be resolved, that the project would soon become unsustainable due to its large size. But of course all of these objections could have been presented in the case of *GNU* and *Linux* and probably were. In order to succeed the project would clearly benefit from being decoupled from any one individual or institution. This is particularly appropriate since, in any case, it can be said to constitute a synthesis and a natural evolution in CAL(L) practice. As an evolutionary framework, its ownership is likely to become more and more diffuse as progressive refinements and departures are experienced through time.

In the final analysis, however, I see the architecture as eminently realistic, and more so than an artificial language like Esperanto or a competing operating system like Linux. The system would run in parallel with existing commercially driven authoring endeavours. Far from competing against them, it might promote some which adopted its end-product creation features. To suggest that this will somehow "kill" all authoring systems is like saying that *Linux* will eventually kill Microsoft

It would certainly encourage the separation of chargeable content and free enablement tools. Furthermore, if we are to exploit the need for authentic materials in language learning courseware, content updates will need to be rapid and frequent, leaving a healthy courseware production industry.

To conclude, DISSEMINATE is in essence a very simple idea, which could be summarised as the implementation in the digital domain of the ringbinder philosophy. It is truly consistent with the digital promise of accessibility, flexibility and fast updates, rather than inertia and redundancy. Is it not a paradox that a piece of courseware, through the long software development cycle, is much more costly in time and finance to produce than a conventional text-book and becomes in effect much more rigid? Is it not the case that individual authors who may have a strong design contribution to make to the field are deterred from doing so by the time and cost factor as well as the range of skills required to produce a professional product? DISSEMINATE combines small and large developments. It combines significant economies of scale associated with large consortia efforts with individual craft and talent, and that alone may give it the dimension of a compelling venture in the eyes of at least a few enthusiasts. The process of incremental, iterative, heuristic courseware construction might be one of the most convincing features of the proposal.

Notes

[1]   I am indebted to John W. Oller Jr., a keynote speaker at the 1996 CALICO conference, for first bringing this fact or rumour to my attention. I have since checked McLuhan's bibliographical references through a library and the Amazon website: http://www.amazon.com. A search at the latter revealed some interesting discrepancies in the use of "massage" as opposed to "message" in both original and posthumous publications. See the bibliographical references which follow these notes (McLuhan & Fiore 1967; McLuhan et al. 1996). A good exposé of the issues of "The medium is the massage" as against "The medium is the message" can be found at the following location: http://www.zverina.com/alternabook/mccluhan.html

[2]   Wida Software began producing simple authoring packages for creating interactive exercises in 1981, beginning with *Teacher's Toolkit* and *Questionmaster*, both of which were designed by Graham Davies. Later packages included Chris Jones's *Choicemaster* and *Gapmaster*, and John Higgins's *Storyboard* and *Pinpoint*. The most recent version of the series of authoring packages by Wida is known as *The Authoring Suite*: http://www.wida.co.uk

[3]   *Fun with Texts* (Camsoft, Maidenhead) was originally produced in 1985 by Graham Davies as a derivative of *CopyWrite* and *Storyboard*. Version 3.0 is now available for Windows: http://www.camsoftpartners.co.uk

[4]   1965 is the date found in most magazines and television programmes which refer to the coining of the term "hypertext". See: Berners-Lee, T. (1996), *The Web: Past, present and future*: http://www.w3.org/People/Berners-Lee

[5]   This is drawn from Levy (1997a:16).

[6]   See Bertin, J-C. (1998) and the chapter by Bertin in this publication.

[7]   Delcloque, P. and Kashko, A. (1997*), Authorlang software prototype* (part of the CALLIFaT Project, University of Abertay Dundee).

[8]   Davies, G. (1997), Private communication: E-mail Message sent to EUROCALL Members Discussion List.

[9]   IBM/Lotus (1997), *Learning Space* courseware production environment, based on *Lotus Notes*. See http://www.lotus.com/engine/branding.nsf/names/learnspace.

[10] WebCT WWW Course Tools. See http://www.webct.com/webct

[11] Auralog (Paris, France), creators of a series of CD-ROMs incorporating speech recognition software, e.g. *Talk to Me, Tell me More*.

[12] Syracuse Language Systems (Syracuse, NY, USA), creators of a series of CD-ROMs incorporating speech recognition software, e.g. *TriplePlay* and *TriplePlay Plus* – now known as *SmartStart*.

[13] A full list of Gagnepain's publications can be found at the Ecole de Rennes website: http://www.uhb.fr/sc_humaines/lirl/pages/niveau1/publitdm.html#anchor512389. I am not aware of any English translation of the work but a brief statement of the Mediation Model or Theory is found on the English part of the LIRL website: http://www.uhb.fr/sc_humaines/lirl/pages/niveau1/lirlenglish.html#anchor761993. Gagnepain (1982) and his other works are a rather "hard read"; a more approachable summary of the theory is given in Bruneau & Balut (1997).

[14] Stallman, R. (1984): The *GNU* website manifesto leading to *Linux*: http://www.GNU.org. I am indebted to Jean-Pierre Messager for first introducing me to *GNU* and *Linux* in late May 1998. A comparison of the original *GNU* manifesto and my main points reveals that there are fewer similarities than I might have imagined, and I agree with a significant number of points made by Stallman. More systematic comparisons between *GNU* and DISSEMINATE will be made in forthcoming publications.

[15] The main European initiative is known as Memorandum of Understanding or MoU. It may be consulted at http://www2.echo.lu/telematics/education/en/news/mou.html Other initiatives share some of the aims of DISSEMINATE: e.g. the MALTED project (http://www.malted.com), *Hot Potatoes* (http://web.uvic.ca/hrd/halfbaked), Ruth Vilmi's *XErcise Engine* (http://www.hut.fi/~rvilmi/XE/Demo). In terms of infrastructure possibilities, it is also worth looking at the practices of the Linguistic Data Consortium: http://www.ldc.upenn.edu

[16] Language Publications Interactive's *Oscar Lake* series of "virtual reality" adventure CD-ROMS, 1997: http://www.languagepub.com

References

Berners-Lee, T. (1996). *The Web: Past, present and future*. Available (2000): http://www.w3.org/People/Berners-Lee.

Bertin, J-C. (1998). L'ordinateur au service de l'apprentissage ou l'apprentissage au service de l'ordinateur? *Les Cahiers de l'APLIUT, XVII.2*, 57–68.

Bickerton, D., Ginet, A., Stenton, T., Temmerman, M. & Vaskari, T. (1997). *Final report of the RAPIDO project*. Plymouth: University of Plymouth (SOCRATES Project TM-LD-1995-1-GB-58).

Bruneau, P. & Balut, P-Y. (1997). *Artistique et archéologie*. Paris: Presses de l'Université de Paris-Sorbonne.

Davies, G.D. (1989). EXPODISC – an interactive videodisc package for learners of Spanish. In E. Buchholz (Ed.), *Fremdsprachenlernen mit Mikrocomputer und anderer Informationstechnik* (pp. 19–22). Rostock: Wilhelm-Pieck University.

Davies, G.D. (1991). EXPODISC – an interactive videodisc package for learners of Spanish. In H. Savolainen & Telenius, J. (Eds), *EUROCALL 91: Proceedings* (pp. 133–139). Helsinki: Helsinki School of Economics.

Felshin, S. (1995). The Athena Language Learning Project NLP System: A Multi-lingual System for Conversation-Based Language Learning. In V.M. Holland, Kaplan, J. & Sams, M. (Eds), *Intelligent language tutors: Theory shaping technology* (pp. 257-272). Mahwah, New Jersey: Lawrence Erlbaum Associates.

Gagnepain, J. (1982, reprinted 1990). *Du vouloir dire: traité d'épistémologie des sciences humaines,* Tome I: *Du signe, de l'outil*. Paris: Livre et Communication.

Hart, R.S. (Ed.) (1981). *The PLATO system and language study*. Special issue of *Studies in Language Learning, 3.1*.

Kenning, M-M. & Kenning, M.J. (1990). *Computers in language learning: Current theory and practice*. New York: Horwood.

Levy, M. (1997a). *CALL: Context and conceptualisation*. Oxford: Oxford University Press.

Levy, M. (1997b). Annotated Bibliography of CALL. In *Annotated bibliography of English Studies*. Lisse: Swets & Zeitlinger.

Lian, A-P. (1988). Distributed learning environments and computer-enhanced language learning (CELL). In J. Dekkers, Griffin, H. & Kempf N. (Eds), *Computer technology serves distance education* (pp. 83–88). Rockhampton: Capricornia Institute.

Marshall, I.M. (1997). *Evaluating courseware development effort: Estimation measures and models*. Unpublished doctoral dissertation, University of Abertay Dundee.

Matthews, C. (1991). *Intelligent CALL (ICALL) bibliography*. Norwich: University of East Anglia.

McLuhan, H.M. (1964). *Understanding media: The extensions of Man*. New York: McGraw Hill. Republished with new introduction by Lapham, L.H., London: Routledge, 1994.

McLuhan, H.M. & Fiore, Q. (1967). *The medium is the massage*. New York: Bantam.

McLuhan, H.M., Fiore, Q. & Agel, J. (1996). *The medium is the massage: An inventory of effects.* San Francisco: HardWired.

Murison-Bowie, S. (1993). TESOL technology: Imposition or opportunity. *TESOL Journal, 3.1,* 6–8.

Murray, J.H. (1995). Lessons learned from the Athena Language Learning Project: Using Natural Language Processing, graphics, Speech Processing, and interactive video for communication-based language learning. In V.M. Holland, Kaplan, J. & Sams, M. (Eds), *Intelligent language tutors: Theory shaping technology* (pp. 243-256). Mahwah, New Jersey: Lawrence Erlbaum Associates.

Schneider, E.W. & Bennion, J.L. (1984). Veni, vidi, vici via videodisc: A simulator for instructional conversations. In D.H. Wyatt (Ed.) *Computer-assisted language instruction* (pp. 41–46). Oxford: Pergamon Press.

6

CALL material structure and learner competence[1]

Jean-Claude Bertin
Université du Havre, FR

1. Introduction

Multimedia language learning materials design is now open to all teachers thanks to the most recent authoring tools which require very limited computer skills. The freedom thus offered designers makes it possible for teachers to meet the needs of all learners even more appropriately. In a recent survey, the author showed a correlation between learners' linguistic competence and their expectations in terms of autonomy. This finding raised the problem of the organisation of browsing in courseware. The aim of this chapter is to present the results of the survey and to suggest a number of paths in order to adapt a given pedagogic content to a variety of expectations.

The interest of multimedia for language learning is no longer contested although its actual efficiency largely depends on the nature of the materials. The most recent and advanced authoring systems now enable teachers to design their own materials without any specific skills in computer programming.[2] Such developments will at last ease teachers' work in meeting learners' needs more accurately, as any given pedagogic content, any "lesson", will readily be made available under different forms to meet varying learning strategies and styles.

Adapting contents to needs means designing learning materials whose structures are appropriate for a range of cognitive attitudes. The question however is to determine the relation between the learners' cognitive profiles and the structure of the pedagogic scenarios. This is the subject of this chapter. The first part will introduce a study carried out among students in the University of Le Havre, France (Annoot and Bertin 1997). This study forms the basis of the theoretical approach which will be outlined before considering more practical applications in a second part.

2. The study

The Le Havre study aimed at defining the conditions of ICT integration in language curricula, taking as a starting point the case of languages for international transport and logistics. It followed a multidisciplinary approach (language teaching and social sciences) consisting in analysing the reactions of several groups of students using computers as part of the specialised language course. The study combined in-course observations and systematic interviews at the end of the academic year. The complete report was published in June 1997 in the University of Le Havre (CIRTAI).[3]

The results mainly pointed to the overall necessity to diversify teaching/learning approaches according to several criteria, one being the ability to work autonomously:

> If autonomous learning situations are to be developed using courseware, they must be conducted by the teacher representing the institution and observing learner behaviour when accessing knowledge. Such an approach needs to be diversified in relation to the degree of learner autonomy. (Annoot & Bertin 1997; my translation)

This first point was further complemented by other observations showing that the efficiency of CALL also varies with each learner's linguistic and communicative competence. These findings, which are further corroborated by Kettemann (1995) and Zähner (1995), enable us to outline the role of learner competence in the appreciation of the CALL environment, as shown in Figure 1.

Figure 1: Learner linguistic competence and learner expectations

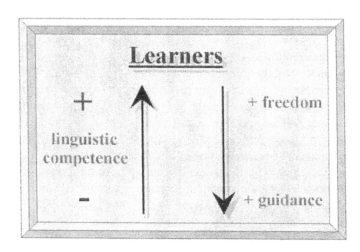

In general, it was noticed that the higher the level of linguistic competence, the more freedom and autonomy is expected both from the specific courseware used and from the overall CALL situation. Similarly, the more learners are aware of their linguistic limitations, the more strongly they feel the need for the teacher's presence. This presence may be either physical, in the multimedia room, or mediated through an explicit form of guidance in terms of courseware selection or pedagogic organisation of individual lessons. The question then arises as to how CALL material designers can expect to meet the variety of needs: how to choose between freedom of hypernavigation and pedagogic guidance?

In the report, I recommended offering the less advanced learners, as a first stage in the CALL curriculum, a preliminary task on specific courseware in order to help them gain confidence before facing tasks requiring a higher level of autonomy. It is important to note that this approach is also adapted to more serialist learners. At the same time, the more advanced and/or holistic learners should be given more abstract tasks resorting to e-mail or the Internet, for example, for direct or indirect communication work. It is hypothesised that at one point or other of this work, these learners will be faced with communication problems that will lead them to turn to more specific activities with appropriate software. This first approach considers the variety of ICT uses as an integral part of strategic competence (Canale & Swain 1980) within the overall CALL situation.

This approach can also be taken when considering the internal structure of hypermedia learning materials specifically. Indeed, the study not only recommended various ways of integrating the different types of computer use, it also pointed to the necessity to adapt each individual multimedia "lesson" to the range of learner expectations and needs. In this new scenario, the first prerequisite is obviously an authoring system which enables rapid material development. (This point will be treated in more detail later in the chapter.) It is then necessary to analyse the conditions of such individualisation of CALL material structure from a theoretical standpoint. Is total freedom of navigation sufficient for language acquisition to take place? If this were true, we could then imagine a technological version of the Natural Approach (Krashen & Terrell 1983): the present craze for the Internet might well indicate that this is sometimes the case, which is one of the reasons which justifies the present chapter. Firstly it is necessary to consider the various aspects of the situation which CALL materials are supposed to generate.

3. The learning process: Organisation and discovery

Cognitive sciences are based on the structuring activity of the brain and try to define systems in which the learner is led to think about what he learns and how he learns (Narcy 1990:267). A "constructivist" approach defines learning as "a process of self-organisation by the learner [which] can only be organised if the learner takes the full responsibility for his own learning" (Wolff 1997:18). It is now generally agreed that a teacher wishing to develop his/her own learning materials should follow an approach close to that defined by Louis Not:

In a perspective that considers knowledge as the product of learner activity, the analysis should consider this activity first, the teacher's activity being defined in a dialectic relation with that of the learner. This is a complete reversal of the classic perspective which considered the teaching act fundamental and the learning act as a natural consequence of teaching. This led to a focus on the teacher, and the obligation for the learner to adapt to him.[4] Instead of leading the learner to adapt to the teaching activity, what is now aimed at is the adaptation of the teacher's activity to that of the learner. (Not 1991:30; my translation).

The other essential notion to bear in mind is heuristics, which leads to a focus on the central function of the discovery of the linguistic material by the learner. It is not however clear whether this discovery should take place in a totally free environment, "Krashen-wise", or within a learning environment that has been first delineated by the teacher. The history of CALL (whatever acronym may have been used at one time or another) has been a shift between two poles corresponding to the two visions outlined above: directed learning and autonomous discovery (Figure 2).

Figure 2: Shifts in CALL material structure

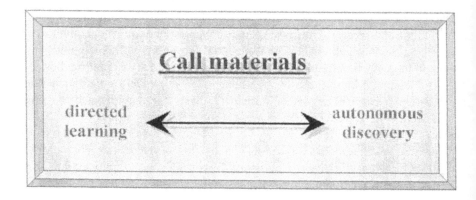

3.1 Directed learning
Organising learning may mean very different things according to the pole one refers to. According to second language learning theories, it was first thought possible to describe the learning process algorithmically. This first stage gave birth in the 1950s to what came to be known as "Programmed Learning", an outcome of Skinner's behaviourism. These theories were then taken a step further by Crowder who proposed teaching "programmes" liable to reduce errors by offering learners more individualised paths. The so-called scientific organisation corresponded to predetermined sequences presented to learners in a linear fashion.[5] The limits of this approach rapidly became clear: individualisation only affected learning pace, but no attention was paid to individual strategies or to the dynamic and pragmatic nature

of linguistic communication. The failure of Programmed Learning is mainly due to the impossibility to define a clearly identifiable algorithm of second language learning which could have been translated into computer language.

Organising learning however may also be conceivable at the level of the presentation of linguistic data, in order to ascertain the relevance of the material in relation to explicit pedagogic objectives. This is especially the case in Language for Specific Purposes (LSP) courses where it is vital to make sure that learners' attention will not be diluted by over-surfing on the multi- or hypermedia systems placed at their disposal. This type of linguistic data organisation aims at maintaining coherence between a teaching "programme" and the general objectives of the course.

Taking into account the results of the above-mentioned study, it is important to note that one should also take into account the need for guidance expressed by the less advanced learners. This should mean offering them access to more clearly structured and organised materials than the Internet or any other hypermedia. The problem is to know how far one can go without reproducing the excesses of Programmed Learning.

3.2 Autonomous discovery
The heuristic approach mainly focuses on the problem of defining sufficiently flexible access to information in order to match the diversity of pace and navigation implied in the process of individual discovery. In the present case, access to the linguistic material should no longer depend on a predetermined navigational pattern but should meet improvised associative criteria meeting the needs arising from a specific situation, thus facilitating the memorisation process. If this process still implies some organisation activity, such activity takes place at the level of the mental processes at work in the learner. The software that simulates – or rather stimulates – language learning should then favour the structuring activity of the learner by offering him access to data with as few limitations as possible. If the literal and caricatural translation of organisation reflects the linear pattern of Programmed Learning, the computer translation of heuristics corresponds to the concepts of hypertext or hypermedia. Hypertext (or hypermedia when data of different natures are combined) may be defined as a document that can be consulted in a non-linear fashion by means of interactive links included in the screen pages. In this context, the Internet is nothing but a huge hypermedia. Complete freedom of navigation combined with a high degree of interactivity has led hypertext to become an indispensable cognitive tool (Wolff 1997) according to the principle that "one only learns what one is trying to learn" (Clergue 1991:264). Ideally hypertext provides an autonomous learning environment which seems to match present second-language learning theories.

Yet, one should beware of adhering only to this pole of our axis (Figure 2) in just the same way as others formerly tried to control and organise too much. One should remember that total autonomy is no longer considered as a means in itself, even if it remains an end to the educational process, which is why the concept has been superseded by that of guided autonomy. Furthermore, the potential of hypermedia (and even more so of the Internet) presents the risk of a loss of reference on the level of linguistic contents – if an institutional curriculum has to be observed – as well as on the level of navigation. As is often the case when dealing with materials design, it is important to take a measured approach and avoid going too far in one direction by privileging technology at the expense of didactics. Lessons should be drawn from the past without totally missing the opportunities presented when new horizons open.

4. The three structural axes of CALL materials

On what bases, then, should the CALL materials designer work in order to try and meet the diversity of his learners' needs in the perspective I have just briefly outlined? In an earlier study (Bertin 1998:62), I developed a theoretical description of CALL materials structure based on three main axes (Figure 3).

Figure 3: The three structural axes of CALL materials

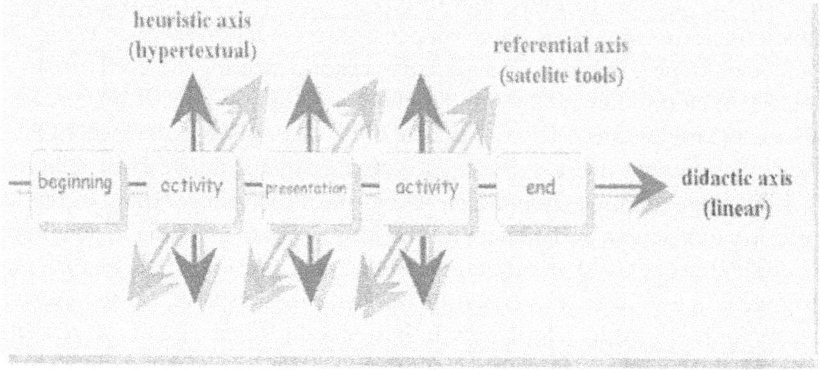

The didactic axis is constituted by the sequence of screen pages designed by the author of the lesson. The heuristic axis appears when the learner is given the opportunity to escape the linear structure, through interactive buttons and other hypermedia links. The referential axis, which is unfortunately not found in all learning environments, implies the availability of complementary or "satellite" tools, whose presence in the virtual learning space favours the learner's structuring activity through questioning, searching for complementary information and selection of data as well as through the growing awareness of the various learning strategies.

If one bears in mind the possibility of combining these three axes by privileging one or the other, it becomes possible to design totally individualised lessons while preserving the objectives and curriculum of the course.

5. Diversifying the approaches

A primary condition to work in such a perspective is to have access to an authoring system that provides a really congenial environment for fast and efficient design or modification of multimedia lessons. Such systems were hard to find until very recently, and few of those that do exist offer both a wide range of didactic activities and ease of use for non computer-literate teachers, i.e. the vast majority of the public who know virtually nothing about programming. A large proportion of the output of researchers consists of finished products that may be convincing on didactic grounds but that require time and expertise totally out of reach of most teachers. Although this is not directly the focus of this chapter, it is necessary to stress the importance of this requirement since the approach suggested here would be totally utopian if one had to spend hours learning specific development techniques to adapt existing multimedia lessons.

Once this point has been made, designing similar pedagogic contents for different learning styles simply means organising the linguistic material and didactic activities along appropriate structural lines.

Figure 4: An example of a linear pedagogic structure

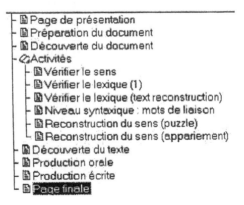

Thus, an important didactic axis (Figure 4) will stress the feeling of the teacher's presence, translated by a stronger linear structure. The different screen pages will correspond to the various acquisition phases recommended by the teacher: inference, discovery, interaction with the linguistic material, evaluation. Conversely, a stress on the heuristic axis will show a systematic resort to hypertextual links and a subsequent reduction in the number of screen pages (Figure 5).

Figure 5: An example of a hypertextual pedagogic structure
(same lesson as Figure 4)

- 🖹 Page de présentation
- 🖹 Préparation du document
- 🖹 Zone d'activités
- 🖹 Découverte du script de la bande son
- 🖹 Page finale

Each screen page will be considered as a suggestion for action, as a starting point (a "node") towards activities or information activated according to individual needs and pace.

These two examples are intended to illustrate the two poles that delineate the teacher/designer's scope. As a general rule, designers now tend to abandon the linear structure to opt definitely for the total freedom of navigation offered by hypertext/hypermedia. Should we then conclude that a strong correlation exists between language acquisition and total freedom? If constraint-free hypernavigation (e.g. surfing on the Internet) certainly favours progress, is this progress really optimal? It could be wise to draw a parallel between navigation in courseware and the importance given to guided autonomy in second language learning theories. Experience shows that one should often temper freedom of navigation with a measure of pedagogic guidance. This recommendation corresponds to the feelings expressed by the learners: whatever their level of linguistic competence or the activity in the multimedia room, learners like to feel the physical presence of the teacher. In any CALL material, the teacher's presence is mediated through the structural organisation of linguistic data and the associated didactic activities. The students interviewed left no doubt on this point. The function of the designer then is to provide the accessories, the setting, which will put the learner on the stage by organising his/her interaction with the language. The teacher will remain hidden in the wings.

The linguistic material of the lesson may be compared with a database. The various modes of consultation will define the learning situation. If the computer is only considered as a mediator between the learner and the target language, however individualised this mediation is enacted (taskless surfing on the Web, for example), the consultation of the database will take place as shown in Figure 6:

Learner's QUESTION => Computer's RESPONSE

The advantage provided by hypertext, even in this case, is the creation of an unorganised informative environment within which the structuring activity of the brain may be activated. However, the learning situation surely differs significantly from the situation outlined here insofar as the ultimate goal is not the mere access to linguistic data – even though this is in itself an element of learning – but second language acquisition.

Figure 6: Consultation of a database

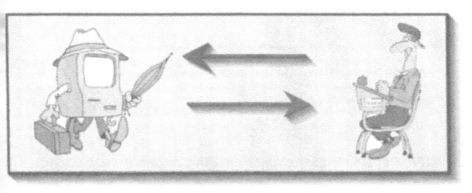

Database :
* learner question
* computer response

In opposition to this database, what I call a "language learning database" is organised along the three above-mentioned axes, as shown in Figure 7.

- Organisation of the linguistic data by the teacher: choice of material in relation to specific or institutional pedagogic objectives. This organisation or guidance, however it is termed, may take on distinct forms which will be described later.
- Problem-solving activity: the communicative function of language implies simulating situations where the learner can read/listen to/watch a document in order to respond (information transfer). In the case of ESP particularly, the two needs analyses I have personally carried out, clearly show that this is a central skill to be developed. In the CALL situation, the computer (seemingly) gives the learner the problem to be solved.
- Finding the answer: when reference and other complementary tools are available for the learner, their use is an important factor of acquisition as their interactive consultation means preliminary analysis on the part of the learner in order to define problems and to organise linguistic data. Such tools have a significant metacognitive function.
- Learner answer + computer evaluation: this is a stage when the learner's hypotheses (learning is a hypothesis-formulating process) are confronted with the correct answers planned by the teacher. This confrontation will take different forms according to the type of activity, of software...
- What is crucial, in this perspective, is the relation between the three components of the CALL situation (teacher – computer – learner), as reflected in the structure of the materials:

Figure 7: Consultation of a language learning database

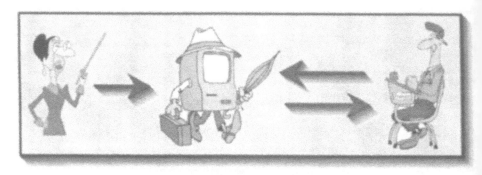

language learning database:
* data organisation (teacher)
* problem-solving activity (computer)
* search for answer (learner)
* submission of answer (learner)
* evaluation (computer)

i   a high degree of interactivity between the learner and the computer;
ii   an organisation of the data by the teacher before the actual CALL situation
     is enacted, taking into account the learners' individual strategies. The idea
     is to provide a network of potential pathways guaranteeing that all learners
     will eventually have navigated through all the phases judged essential,
     whatever the form and nature of their navigation. This may imply design-
     ing distinct materials with a similar objective, providing varying degrees of
     guidance and freedom;
iii organisation of nodal points from which hypernavigation will take place in
     strict observance of pedagogic objectives.
iv   While the learner remains at the centre of the pedagogic situation and en-
     joys total navigational freedom within the learning database, the teacher
     continues to pull the strings as a stage director. Thus, the learner in front of
     the computer will not be the passive spectator of a show designed for his
     own use: he will also be an actor, in the way contemporary playwrights
     sometimes involve the spectator in the action. Likewise, the play enacted
     by the computer is not immutable: only the prompts are provided, but the
     spectator/actor's place in the script has been planned – without knowing
     exactly his exact degree of participation nor the nature of this participation.
     For the play to take place, its author (the teacher) should have provided
     guidelines that will somewhat limit the freedom of the spectator/actor without
     rigidly fixing his character.

6. Different forms of guidance

What means are available for the teacher in order to fulfil this guiding function within a learning hypermedia database? The teacher's role does not disappear behind the heuristic axis in CALL materials. His guiding function is exercised mainly through the structural choices he makes when designing materials, by combining the didactic, heuristic and referential axes, as noted by Robin Goodfellow and Peter Metcalfe: "The dangers of being 'lost in hyperspace' are avoided by the availability of a linear anchor and careful layering of information" (Goodfellow & Metcalfe 1997:6). Pedagogic guidance may take on an explicit or implicit form, according to the axis chosen by the teacher/designer. An explicit form of guidance will rather satisfy less advanced learners while implicit guidance will suggest guidelines without harming more advanced ones.

6.1 Guidance through the didactic (linear) axis.

This is the most explicit and constraining, not to say elitist (Moro 1997:73), form of guidance if it is not complemented by ample navigation facilities. The pedagogic organisation of the material is divided into screen pages reflecting the various stages through which the teacher wants the learner to pass. Linear navigation is commonly controlled through a graphic interface reproducing well-known CD players or VDUs (Figure 8).

Figure 8: The linear controller of the "Learning Labs" environment

The resort to the didactic axis corresponds both to this "linear anchor" mentioned by Goodfellow and Metcalfe, whose function it is to avoid being lost in hypernavigation, and to the desire to ascertain that the learner's activity will fit into the curriculum. It is important to stress again, however, the importance of not overdoing things by devising too strong a linear axis which would limit the potential evolution of learning strategies that are never fixed (Narcy 1990:103) and whose consequence would be a standardisation of the learning process (Moro 1997:73). The environment I use includes the possibility for the learner to get out of the navigation scheme suggested by the teacher, via an interactive map of the material (Figure 9). A simple combination of keys[6] makes it possible to directly access the desired point of the material, to skip a phase thought inappropriate, to go back in the lesson to check a specific item.

This linear anchor seems essential to me, even in the case of materials making a large use of hypertext. I fully agree with Graham Davies on this point:

A certain amount of browsing is to be encouraged, and I firmly believe in allowing the learner to follow his/her own inclinations as far as possible.

Software, however, needs to offer what has become known, in CALL jargon as the "default route". Diana Laurillard puts the argument succinctly as follows: "a default route is the route through the material that the author believes to be optimal. Completely open-ended program structure can make students anxious – they like to know what they are supposed to do. It must always be possible to deviate from the default route, but it should be clear what it is, so that they can just follow it through. This saves students having to make decisions at every turn, and may also encourage them to consolidate rather than keep moving on". (Laurillard 1993:2) (Davies 1997:43)

Figure 9: An example of an interactive map

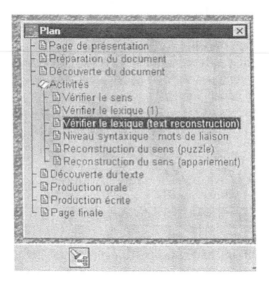

6.2 Guidance through the heuristic (hypermedia) axis

At first sight, resorting intensively to the heuristic axis would seem to contradict the notion of guidance, since the objective is navigational freedom within the learning database in order to leave the learner structure his/her interlanguage. Yet, I have already hinted at the need to give the learner the opportunity to visualise the complete database in order to plan his/her navigational decisions and to enable him/her to get back to the original nodal point. This is the second function which can be attributed to the interactive map (Figure 9). Besides, if the material's structure is fundamentally based on hypernavigation, the latter depends essentially on the nodal points that have been devised when constituting the learning database. Goodfellow and Metcalfe insist on the emphasis placed on the "careful layering of information".[7] Thus, hypermedia designers are commonly advised not to offer more than three levels of information. This is an example of implicit guidance, since such data organisation limits the effective capacity of the learner to surf from one item to

.nother. In the same way, the designer of a hyperdocument may opt between two types of presentation of navigational choices, when planning important nodal points.

Providing explicit guidance will mean organising the database into chapters and sub-chapters, as in a book. These constitute hierarchical choices that naturally lead the learner to discover the document in the suggested order, without any contraint, however (Figure 10). In our study, this form of guidance proved to meet learners' expectations independently of their levels of competence since it associates the teacher's implicit presence and the learner's freedom of choice.

Figure 10: An example of hierarchical choices (hyperdocument)[8]

Conversely, the choices offered to the learner may be non-hierarchical, as in the case of lists of textual and graphical elements, or of interactive buttons (Figure 11).

Experience shows that this alternative, offering unrestricted navigational freedom, proves mainly efficient for learners capable of autonomous decisions, which is unfortunately not the general case. It is common practice for learners faced with this type of non-hierarchical choice to request the teacher's direct intervention!

6.3 Guidance through the referential axis (satellite tools)

The referential axis may be conceived as a variation on the heuristic axis insofar as it corresponds to the learner's decision to activate or not the tools provided in the virtual learning environment: dictionaries, grammatical reference, databases or access to websites... The availability of such software tools is in itself an implicit form of guidance not only at the level of the discovery of the linguistic material, but also at the level of the development of procedural knowledge. "Help can be provided in different ways, for example by making the learner acquainted with learning strategies or with learning tools" (Wolff 1997:19). The provision of a referential dimension in CALL materials is a notion I have been experimenting since the early

1990s and which gave birth to my concept of a "Learning Space".⁹ Complementing hyperdocuments with a set of reference tools is meant to compensate for the loss in learning resulting from a constraint-free navigation. The mere presence on the screen of their icons constitutes a form of suggestion, or attraction: indeed, the observation of "hyperlearners" shows that, when hyperlinks are clearly identifiable on the screen, they tend to explore each of them successively. The consultation of similar traditional reference documents, even if easily available, proves to be much rarer.

*Figure 11: An example of non-hierarchical choices (hyperdocument)*¹⁰

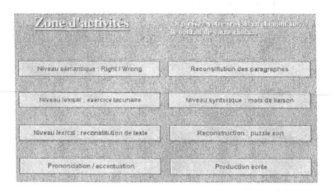

My personal experience is based on the use of the following tools (Figure 12):

- a grammatical reference database, *Brush up your Grammar!,* designed on the basis of the learners' most frequent mistakes;
- an English irregular verbs reference database;
- a specialised dictionary designed by the teacher (e.g.: English for international transport);
- a general English dictionary linked to the former by hypertext: when a term is not referenced in the specialised dictionary, it is automatically searched in the general English dictionary provided in the environment;
- encyclopaedias;
- a personal dictionary created by the learner and printable on paper. The presence of this dictionary will lead the learner to organise, then select the data he/she wants to save, thus activating structuring and memorising processes.
- access to the Internet.

Figure 12: Referential axis – the satellite tools icon bar from "Learning Labs"

6.4 Summary: use of the various navigational modes (guidance and freedom)

Figure 13: Navigational modes and degree of guidance / freedom

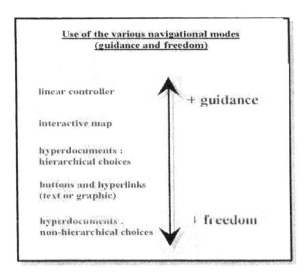

7. Conclusion: an example of mixed structure

Freedom or guidance? On principle, one might say with Bernard Moro that the choice is open between two perspectives:

- a traditional linear perspective based on Cartesian rigour and objective-oriented;
- a more "enjoy-the-ride" perspective based on an intensive resort to hypernavigation (Moro 1997:74).

None of these perspectives used on its own, however, seems to really answer learners' expectations. As this chapter has tried to show, language teachers have now access to a range of tools that they can use in order to diversify the pedagogic structure of the materials they use or design and to match their learners' needs and levels of competence as closely as possible. With a minimum degree of experience, they will rapidly outline a few basic structures corresponding to the main objectives they have set themselves. Adapting CALL materials should rarely take longer than half an hour or so. If the authoring system used deserves its name, only the basic structure will have to be altered, as well as a few screen layouts. The essential elements of the materials (texts, pictures, sounds, exercises, video clips...) will seldom require significant alterations.

As an example, and as a conclusion to this paper, here is one of the basic structures I use for the development of listening comprehension skills (Figure 14). The didactic axis is constituted by the main guideline defined thanks to the preliminary needs analysis: oral comprehension, in English for specific purposes, is commonly one of the components of a more global skill, information transfer. The linear structure of this lesson corresponds to a now generally accepted pattern of the communicative approach: preparation and inference phase, task-oriented discovery of the document, comprehension check and development of linguistic competence, oral production work, reinforcement by use of new knowledge in a different context (written production task, in this example). This is the default route alluded to previously.

Figure 14: An example of mixed structure
(same lesson as Figures 4 and 5)

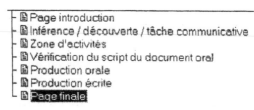

The heuristic axis remains strong, since phases 2 and 3 correspond to non-hierarchical choices (see Figure 11). The permanent availability of the interactive map icon reinforces the freedom of navigation offered to the learner. Phases 4 (access to script of the document) and 6 (written production) implicitly lead the learner to resort to the satellite tools (referential axis). The integration of these three axes constitutes my concept of a "learning space" (Figure 15).

Figure 15: The three axes of the learning space

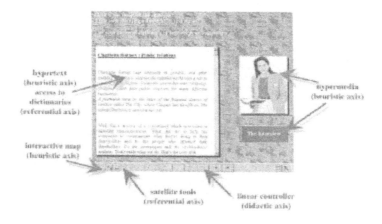

Notes

[1]   This article was originally published in 1998 as "Conception de leçons multimédia: liberté ou guidage?" *Asp., 19–22*, 313–31.

[2]   The authoring system that I use and which served as a basis for this chapter is called *Learning Labs*.

[3]   Centre Interdisciplinaire de Recherche en Transport et Affaires Internationales, 25 rue P. Lebon, F-76600 Le Havre.

[4]   This still appears to be the case with a significant number of sofware packages, which turn out to be mere technological transcriptions of former pedagogic methods.

[5]   Crowder somewhat refined the model by offering a tree structure organisation. Navigation in the structure nevertheless remained essentially linear.

[6]   [CTRL] + click

[7]   See also McAleese & Green 1990.

[8]   This example is an extract from *Brush up your grammar!*, a hypertext-based grammar consultation package.

[9]   This example is an extract from my CD-ROM *English for International Transport and Logistics*.

[10] This original concept, renamed *Learning Labs* since then, should not be confused with the software of the same name distributed by the American firm Lotus. For further reference concerning this concept, see Bertin 1993a and 1993b.

References

Annoot, E. & Bertin, J-C. (1997). *L'Intégration des Nouvelles Technologies dans les formations en Langue de Spécialité (cas des langues du transport et de la logistique)*. Rapport du CIRTAI, Le Havre: Université du Havre.

Bertin, J-C. (1993a). *Learning Space: The NeXT concept?* Paper presented at the Second Conference of ESSE (European Society for the Study of English). Bordeaux, September 1993.

Bertin, J-C. (1993b). New concepts and new machines for CALL: The NeXT Learning Space. *Computer Assisted Language Learning, 6*, 203–213.

Bertin, J-C. (1998). L'ordinateur au service de l'apprentissage ou l'apprentissage au service de l'ordinateur? *Les Cahiers de l'APLIUT, XVII.2*, 57–68.

Canale, M. & Swain, M. (1980). Theoretical bases of communicative approaches to second-language teaching and testing. *Applied Linguistics, 1.1*, 1–47.

Clergue, G. (1991). *Les Ateliers de la Connaissance*. Paris: INJEP, Dossiers Pédagogiques.

Davies, G. (1997). Lessons from the past, lessons for the future: 20 years of CALL. In A-K. Korsvold & Rüschoff, B. (Eds), *New technologies in language learning and teaching* (pp. 27–51). Strasbourg: Council of Europe.

Goodfellow, R. & Metcalfe, P. (1997). The challenge: Back to basics or brave new world? *ReCALL, 9.2*, 4-7.

Kettemann, B. (1995). How effective is CALL in ELT? *ReCALL, 7.1*, 49–53.

Korsvold, A-K. & Rüschoff, B. (Eds) (1997). *New technologies in language learning and teaching.* Strasbourg: Council of Europe.

Krashen, S. & Terrell, T. (1983). *The natural approach: Language acquisition in the classroom.* Oxford: Pergamon.

Laurillard, D. (1993). *Program design principles.* Hull: University of Hull, TELL Consortium, CTI, Centre for Modern Languages.

McAleese, R. & Green, C. (Eds) (1990). *Hypertext, state of the art.* Oxford: Intellect Books.

Moro, B. (1997). A pedagogy of the hypermedia. In A-K. Korsvold & Rüschoff, B. (Eds), *New technologies in language learning and teaching* (pp. 69–78). Strasbourg: Council of Europe.

Narcy, J-P. (1990). *Apprendre une langue étrangère – didactique des langues: Le cas de l'anglais.* Paris: Les Editions d'Organisation.

Not, L. (1987; 1991). *Enseigner et faire apprendre. Eléments de psychodidactique générale.* Toulouse: Privat.

Wolff, D. (1997). Computers as cognitive tools in the language classroom. In A-K. Korsvold & Rüschoff, B. (Eds), *New technologies in language learning and teaching* (pp. 17–26). Strasbourg: Council of Europe.

Zähner, C. (1995). Second language acquisition and the computer: Variations in second language acquisition. *ReCALL, 7.1*, 34–48.

CALL software

Learning Labs – learning environment and multimedia authoring system.

Brush up your grammar! – a hypertext-based grammar reference package.

ATIL (anglais du transport international et de la logistique) / English for International Transport and Logistics.

The above materials are distributed by Learning Labs, Le Vaisseau, 120, Boulevard Amiral Mouchez, 76087 Le Havre Cedex, France:
htpp//ourworld.compuserve.com/homepages/LearningLabs
E-mail: LearningLabs@compuserve.com

7

From gap-filling to filling the gap: A re-assessment of Natural Language Processing in CALL

Sake Jager
Rijksuniversiteit Groningen, NL

1. Introduction

Just as the preparation of this paper began, three press reports relevant to its subject appeared in the papers almost simultaneously:

i   Dutch Railway Systems had added speech technology to their computer-ised travel information system to make it possible to obtain detailed travel information without the intervention of a human operator.
ii Dutch Telecom would be supplying voice-operated directory services based on speech processing before the end of the year; a service which had been available for some time in cellular systems.
iii A team of British researchers had won a contest to create a computer pro-gram which holds a conversation so naturally it is difficult to distinguish it from a human.

Since Natural Language Processing (NLP) is finally beginning to make its mark in commercial applications, and researchers are starting to build systems which can pass the Turing test, it is time to re-assess the role of NLP in CALL, where it has lived a somewhat precarious existence. A recent conference on the subject organ-ised by the University of Groningen showed that there is considerable interest among language teachers, NLP researchers and publishers to redefine the ways in which NLP can be used in CALL (cf. Jager et al. 1998). This chapter will examine the impact of NLP on CALL from a double perspective: what it has to offer and what language learning requires. It will be argued that the application of NLP in CALL is constrained by several parameters in the didactic setting, including learning style, learning task, didactic method, linguistic domain, student competence, and the role of the teacher. Rather than diminishing the potential for NLP in CALL, a keen awareness of these constraints will help to integrate NLP technologies into CALL, both as tools and as embedded systems.

2. Defining the role of NLP in CALL

Natural Language Processing is a branch of Artificial Intelligence (AI) concerned with the computational modelling of the perception, analysis, interpretation, and generation of natural language (NL). NLP techniques have been applied success-fully for some time in NL database querying systems. Although more advanced systems are available in experimental settings, the commercial deployment of NLP to date typically involves restricted application tasks and limited linguistic domains (cf. Bates & Weischedel 1993; Roe & Wilpon 1994; Pereira & Grosz 1994). The use of NLP in CALL is traditionally associated with Intelligent CALL (ICALL). Several forms of ICALL fall under the more general category of Intelligent Tutor-ing Systems, an offshoot of AI aimed at modelling the cognitive processes involved in learning and teaching aspects of a certain domain. A distinguishing characteristic of many ITSs (not only of those used for language learning; cf. Kearsley 1987) is the use of natural language in the communication between user and machine. Like many other applications of NLP, linguistic coverage of these systems is highly re-stricted, also because the provision of tutor and student modules acting upon the domain in intelligent ways adds to the complexity of the program. In spite of suc-cesses in some areas (e.g. Bull 1994), ICALL has not really come of age within the CALL community as a whole. Even the most concerted attempts to involve re-searchers, teachers, publishers, and industry, such as the EC-funded ReCALL-project[1] have not been continued. Until now channelling NLP into CALL by way of ICALL has had little success.

Mainstream CALL, riding on the back of general technological innovation and gradually adapting to changing pedagogic-didactic insights, has fared very well without the use of NLP. With its emphasis on authentic materials, learner autonomy, and investigative and collaborative learning, modern CALL is particularly well served by hypermedia and communication technologies, such as the WWW, e-mail, chat-ting and videoconferencing. The use of more specialised technologies, such as NLP, is at best regarded as complementary to these general technologies. Even then, the question should be asked whether alternative means of achieving the same task already exist or could be devised. A case in point are the traditional activity types, such as multiple choice, matching and gap-filling. Although their relative impor-tance has decreased with the rise of open-ended activity types, they continue to be used to provide a specific focus within functional, structural or cultural aspects of language learning. Of these, the gap-filling type may be regarded as a trivial form of NLP, which can be implemented by a simple pattern matching mechanism com-paring student input to entries in a list of possible answers. Although listing all possible answers can be quite cumbersome, using this non-intelligent method has often proved more efficient (in terms of cost) and more effective (in terms of coverage) than deploying full-scale NLP techniques, such as morphological analysis and parsing. To determine the potential of NLP in CALL, one should therefore primarily target applications where other techniques are unavailable or undesirable.

3. NLP: Reading and writing skills for training

Open-ended activity types, such as reading and writing tasks, form one such target. The use of NLP in support of reading and writing has great potential, especially if the role of the independent, emancipated learner is acknowledged. NLP techniques, such as lemmatisation and part-of-speech disambiguation, can be used to assist learners studying authentic text materials who want to explore the meaning and use of certain words and phrases in (one or more) dictionaries or corpora at the same time. NLP-enhanced programs offering these kinds of functionality are currently available (Roosma & Prószéky 1998; Dokter & Nerbonne 1998). In the same fashion, grammar and style checkers may serve to assist the learners in writing. Since word processing programs are increasingly used as e-mail editors, the use of such tools in computer-mediated communication, one of the key areas in CALL today, is possible. Although the main thrust in developing such programs comes from commercial applications targeting a business community and not always up to the task yet, alternative products specifically geared to language learning audiences are now beginning to appear on the market. An instance of this is NativeEnglish Writing Assistant, a commercial program from INSO Corporation which plugs into the major word processing programs and Internet browsers. In addition to bilingual dictionaries, the program identifies common errors in spelling, grammar and style to German, Spanish and French learners of English.

It is well known that these programs are far from perfect yet, but they can be a valuable asset in any language training programme. In view of the present shortcomings, there are obvious opportunities for NLP to develop these programs further. Fortunately for NLP, there is no need to wait until 100% accuracy of such programs has been achieved, for the very language learning tradition which supports the use of such tools in learning also emphasises the role of the autonomous learner who is capable of either accepting a piece of advice or putting it aside. Offering advice, suggesting alternative options, and making students think: in the spirit of Higgins' "pedagogue" approach (Higgins 1983), this should be the aim of NLP in the tools arena. It is especially within this context that NLP may be employed as a useful tool in open-ended activity types. Thus the difficult, probably unattainable, task of achieving fully reliable, unambiguous analysis of user input is alleviated by pedagogic-didactic considerations which make the successful application of NLP possible today.

4. NLP for training in speaking skills

The greatest potential for NLP in CALL today is in processing spoken language. Speech is a core modality in language use. The general availability of sound-enabled multimedia applications has contributed to establishing the position of spoken language in CALL. But while the integration of audio, video and text in one program is perfectly suited to provide the authenticity and immersion settings essential for language learning, student interaction with the program is usually limited to the traditional activity types discussed above and simple listen-and-repeat exercises. Although traditional, non-NLP techniques may prove sufficient in training receptive skills, they are of limited value in training production skills. Before examining this issue in more detail, the present state of the art in speech processing (SP) will be briefly reviewed.

Following Cohen & Oviatt (1994), the field of speech processing may be broken down along different dimensions. The first, rather obvious distinction, is between speech analysis and speech synthesis. In speech analysis a further distinction must be made between speech recognition, i.e. identifying the words being spoken, and speech understanding, i.e. interpreting the intended meaning of what is being said. Cohen & Oviatt note that understanding in this sense is possible only with highly constrained application tasks (e.g. to retrieve information from a database) and in restricted discourse domains. The electronic information systems used as examples at the beginning of this chapter are instances of natural language understanding systems. These systems also comprise a speech synthesis component.

Speech synthesis comes in three different forms (Cohen & Oviatt 1994:38): digitised speech, i.e. assembling and playing back human speech; text-to-speech: the rendering of arbitrary text to speech on the basis of a set of pronunciation rules; concept-to-speech: the use of speech in systems which are conceived of as fully conversant partners making appropriate decisions independently and contributing to dialogues spontaneously and accurately. This is the archetypal view of the speaking computer, known from SF films, but pursued seriously by major research institutes today under the collective name of intelligent agents.

Critical factors for the successful application of speech technology include the following (based on Schafer 1994):

- Vocabulary size and content: Successful application is easiest with a limited vocabulary consisting of unambiguous words.
- Fluent speech vs. isolated words: Words spoken in isolation are easier to recognise than utterances of continuous speech.
- Constraints on grammar and speaking style: A better success rate is possible if the range of permitted structures is constrained and a standard speaking style can be assumed.
- The need for training of the recognition system: Adaptation of a system for a small set of speakers is easier than for larger sets.
- The quality and naturalness of synthetic voice response: The conditions of

use may determine the kind of synthesis used; if high quality and natural speech are required, digitised speech may be the best option, although implementing suprasegmental features such as intonation remains extremely difficult.

- The way the system handles its errors in speech understanding: A commonly applied technique is to ask the user for confirmation.
- The availability and convenience of alternate communication: Relevant in two respects: Does the system fill a gap as a primary means of communication? Can the system resort to alternative modes if the system fails altogether?

All of these factors, which Schafer defined for general-purpose speech processing, are highly relevant for CALL. To appreciate the contribution that speech technology can make to CALL, Pennington & Esling (1996) distinguish between "mechanical" and "meaningful" aspects of spoken language competence and this distinction is applied in the discussion below.

4.1 Mechanical aspects of spoken language
The "mechanical" aspect of spoken language "involves learning to discriminate and produce the sounds of a language and tie these together in fluent strings of sounds" to form complete utterances. The "meaningful" aspect "involves learning to build grammatically coherent utterances and to tie these to communicative functions" (Pennington & Esling 1996:154).

In the mechanical dimension, there is a distinct need for applications for pronunciation training. Pennington & Esling (1996), Jones (1997) and Delcloque (1995) argue for contextualised pronunciation training on the basis of continuous stretches of text (not just mimicry on the basis of discrete sounds). They all point to the need for feedback on student input. And while the traditional non-NLP activity types can be used to provide feedback in training discrimination skills, they are of little utility in providing feedback on pronunciation.

Pennington & Esling (1996) describe several applications for suprasegmental and segmental phonology. The programs described (e.g. SPELL and Visi-Pitch) invariably offer visual displays of spectral characteristics (waveforms, pitch, intensity and duration, individual phonemes, etc.) A common practice activity, if there is one, consists of offering a set of targets which speakers have to try to match. Needless to say, these programs have several drawbacks, some of which are pointed out by Pennington & Esling.

Users may find it difficult to relate their own input to the visual representation, and the reliance on phonological properties makes the programs less suited to students having no background in phonology. There is thus a definite need to provide accurate but simplified feedback to which beginning learners will find it easy to relate. An interesting avenue in this direction is taken by Carson-Berndsen (1998), who describes a system offering animated displays of articulatory movements on the basis of real-time input. A figure generator generates the positions displaying them as animations in the vocal tract, which students should not find too hard to interpret. An interesting aspect of this application is that it does not require a vocabulary listing to work, since the system has been implemented on the basis of the phonotactic rules for German (the grammar for permissible combinations of phonemes in a particular language).

The systems referred to above leave it up to the student to decide on the accuracy of pronunciation. And although such an approach may be defended on pedagogic grounds, students may not always be able to form an adequate judgement of their own pronunciation (for a detailed discussion of these aspects, see Jones 1997). Witt & Young (1998) take the feedback one step further by setting up a pronunciation scoring algorithm, calculating scores at the phonemic level in stretches of running text. The pronunciation score is defined as the ratio of the likelihood of the phoneme which should have been said and the likelihood of the phoneme that actually has been said. Experimental results of this so-called Goodness-of-Pronunciation (GOP) method show that it allows the system to tell the student which phonemes were pronounced incorrectly, and which phonemes were used instead of the correct one. Unlike the system described above, the GOP method requires a pronunciation dictionary to operate. Interestingly however, it offers a mechanism for adjusting the recognition threshold to increase or decrease the acceptance of certain sounds.

4.2 Meaningful aspects of spoken language

The teaching of pronunciation is best seen as a step towards more meaningful, communicative practice (Jones 1997; Pennington & Esling 1996). Speech recognition is increasingly used for applications in the "meaningful" dimension. One such application is the use of automatic recognition for vocabulary training. Recognition is relatively easy, although by no means simple, since the lexicon is constrained and words can be offered as isolated units. This makes this type of application useful for beginning learners. A major advantage of the spoken input compared to written input here is the possibility of dissociating pronunciation from the orthographic representation of words. (One such system, for example, is TriplePlay Plus from Syracuse Language Systems.)

More challenging however are applications which attempt to implement speech recognition in scenario-based discourse activities. The limitations of the ubiquitous listen/view-and-repeat activities in multimedia CALL are felt particularly sharply in this field, where student interaction with the program is usually restricted to record-and-play-back, with little room for variation in the input and no possibilities for feedback. Bernstein (1997) describes a system which holds promise for breaking this pattern and taking discourse-based activities beyond the simple listen-and-repeat format. The Subarashii system, an experimental program for learning Japanese, has been specifically designed to understand what a student is saying in Japanese and to respond in a meaningful way in Japanese. The instructional design is based on "seemingly open-ended dialogues" which are implemented in the form of encounters. Only student directions are in written English, any communication in the context of the encounters is in spoken Japanese. Each encounter consists of a script defining possible pathways on the basis of students' anticipated responses. The program knows how to deal with a large set of possible responses, because the data for each encounter have been collected in a prototyping environment automatically recording every student input together with its frequency. Although the Subarashii system may in some respects be reminiscent of the often discredited [CALL tradition, it clearly heralds a new phase in truly intelligent CALL by its close attention to didactic-pedagogic design, its reliance on high-tech as well as low-tech tools, and its plea for integration with traditional activity types to make it possible to focus on specific aspects of language learning.

In sum then, how does SP-enhanced CALL fare in the light of the critical factors for general-purpose SP listed earlier? Given that the discourse domain may be sufficiently constrained to make useful applications of speech processing in CALL possible, the critical factor most relevant for CALL is how it will be able to handle speaker errors. The system has to be sufficiently robust to deal with non-native speech, which may have anomalies in pronunciation, vocabulary and grammar. CALL applications in this field are different from database applications referred to earlier, since the objective is to give a qualitative judgement on the input rather than extract information. A CALL application for pronunciation cannot resort to simple verbal strategies, like "Could you say that again please?", at least not indefinitely. In addition, there is no "definite quantifiable standard" (Pennington & Esling 1996:173) of acceptability for pronunciation. The indeterminacy as to what is to be regarded as a deviation is one of the hardest problems to solve for speech recognition to work. It crucially depends on the definition of a suitable target model in language teaching, which is equally elusive (cf. Jones 1997). Bernstein (1997) points out that:

> [t]he key to the future of multimedia aided language learning systems will be their ability to understand and judge continuous spoken language with programmable levels of acceptance. (Bernstein 1997)

This requires "an adjustable threshold of rejection".

5. Conclusion

Defining what is acceptable and what is not is obviously not a task for language engineers alone. In order to make real advances in this area, and this applies to NLP in CALL in general, language engineers, language teachers and course designers have to work closely together to determine whether what is technologically feasible is also pedagogically desirable. This chapter has highlighted some areas where LNLP may be introduced, but it is important to place the use of NLP techniques in the larger perspective of language learning. Regarding the computer as the sole interlocutor in the language learning process would be a serious flaw in educational design. Equally important, perhaps more important, is its role in putting language learners in touch with each other, with teachers and with native speakers. In a recent study, Salaberry (1996) argues for the benefits of computer-mediated communication in particular. His basic argument that NLP had better be dispensed with, since complete coverage of the language will never be possible, seems unfounded however. There are several gaps in current CALL practice that NLP can fill, provided it is not regarded as an overall solution for CALL. It should be left to all those involved in CALL to define where the intersection between pedagogy and technology will be most useful.

Before real implementation is possible, more research is necessary in the area of systems integration. Many NLP applications require hardware platforms not available to the general language learning community. The general availability of the WWW (which we have already seen to be crucial with regard to achieving pedagogic-didactic objectives, such as communication and exploration), together with platform-independent programming languages will make it a lot easier to fulfil this requirement. Interestingly, both the programs described by Bernstein and Carson-Berndsen are built using these technologies. The Web is important in other respects too. Co-operative projects can be set up using Web technologies, and the distribution of programs to end users will be facilitated. The Web will also provide access to a large user-base, which allows for the new methods of data collection and large-scale effectiveness studies. Developing NLP-enhanced applications for the Web is clearly one of the most promising ways to further general acceptance of NLP as a working technology in CALL.

Notes

[1] ReCALL: Repairing Errors in Computer-Aided Language Learning. *Language Engineering: Progress and prospects '98* (pp. 70–71). Luxembourg: European Commission Telematics Applications Programme, DGXIII, Telecommunications, information market and exploitation of research. See also:
http://www.infj.ulst.ac.uk/~recall
http://guagua.echo.lu/langeng/projects/recall

References

Bates, M. & Weischedel, R.M. (Eds) (1993). *Challenges in Natural Language Processing*. Cambridge: Cambridge University Press.

Bernstein, J. (1997, April). *Speech recognition in Japanese spoken language education*. Paper presented at the conference on Language Teaching and Language Technology, University of Groningen.

Bull, S. (1994). Student modelling for Second Language Acquisition. *Computers & Education, 23.1/2*, 13–20.

Carson-Berndsen, J. (1998). Computational autosegmental phonology in pronunciation teaching. In S. Jager, Nerbonne J. & van Essen A. (Eds), *Language teaching and language technology* (pp. 11–20). Lisse: Swets & Zeitlinger.

Cohen, P.R. & Oviatt, S.L. (1994). The role of voice in human-machine communication. In D.B. Roe & Wilpon, J.G. (Eds), *Voice communication between humans and machines* (pp. 34–75). Washington DC: National Academy of Sciences.

Delcloque, P. (1995). The design of a French pronunciation tutor. In B. Rüschoff & D. Wolff (Eds), *CALL and TELL in theory and practice: The proceedings of EUROCALL 1994* (pp. 98–109). Hull: EuroCALL.

Dokter, D. & Nerbonne, J. (1998). A session with Glosser-RuG. In S. Jager, Nerbonne J. & van Essen A. (Eds), *Language teaching and language technology* (pp. 88–94). Lisse: Swets & Zeitlinger.

Higgins, J. (1983). Can computers teach? *CALICO Journal, 1.2*, 4–6.

Jager, S., Nerbonne, J. & van Essen, A. (Eds) (1998). *Language teaching and language technology*. Lisse: Swets & Zeitlinger.

Jones, R.H. (1997). Beyond "Listen and repeat": Pronunciation teaching materials and theories of Second Language Acquisition. *System, 25.1*, 103–112.

Kearsley, G.P. (Ed.) (1987). *Artificial Intelligence & instruction: Applications and methods*. Reading MA: Addison-Wesley.

Pennington, M.C. & Esling, J.H. (1996). Computer-assisted development of spoken language skills. In M.C. Pennington (Ed.), *The power of CALL* (pp. 153–189). Houston: Athelstan Publications.

Pereira, C.N. & Grosz, B.J. (Eds) (1994). *Natural Language Processing*. Cambridge, MA: MIT Press.

Roe, D.B. & Wilpon, J.G. (Eds) (1994). *Voice communication between humans and machines*. Washington DC: National Academy of Sciences.

Roosmaa, T. & Prószéky, G. (1998). GLOSSER – Using language technology tools for reading texts in a foreign language. In S. Jager, Nerbonne J. & van Essen A. (Eds), *Language teaching and language technology* (pp. 101–107). Lisse: Swets & Zeitlinger.

Salaberry, M.R. (1996). A theoretical foundation for the development of pedagogical tasks in computer mediated communication. *CALICO Journal, 14.1*, 5–34.

Schafer, R.W. (1994). Scientific bases of human-machine communication by voice. In D. B. Roe & Wilpon, J. G. (Eds), *Voice communication between humans and machines* (pp. 15–33). Washington DC: National Academy of Sciences.

Witt, S. & Young, S. (1998). Computer-assisted pronunciation teaching based on automatic speech recognition. In S. Jager, Nerbonne J. & van Essen A. (Eds), *Language teaching and language technology* (pp. 25–35). Lisse: Swets & Zeitlinger.

8

Human Language Technologies in Computer-Assisted Language Learning

Mathias Schulze
UMIST, Manchester, UK

1. Introduction

> ... there is no doubt that the development of tools (technology) depends on language – it is difficult to imagine how any tool – from a chisel to a CAT scanner – could be built without communication, without language. What is less obvious is that the development and the evolution of language – its effectiveness in communicating faster, with more people, and with greater clarity – depends more and more on sophisticated tools (*Language and technology: From the Tower of Babel to the Global Village* 1996:1).

The publication from which the above quotation is taken includes the following examples of language technology (using an admittedly broad understanding of the term):

- typewriter
- ballpoint pen
- spell checker
- word processor
- grammar style / style checker
- thesaurus
- terminology database
- printing
- photocopier
- laser printer
- fax machine
- desktop publishing
- scanner modem
- electronic mail

- machine translation
- translator's workbench
- tape recorder database search engines
- telephone

(op. cit. pp. 2–25)

Many of these are already being used in language learning and teaching. Today most of the research and development that aims to enable humans to communicate more effectively with each other (e.g. e-mail and Web-conferencing) and with machines (e.g. machine translation and natural language interfaces for search engines) is done in Human Language Technologies (HLT):[1]

> The field of human language technology covers a broad range of activities with the eventual goal of enabling people to communicate with machines using natural communication skills. Research and development activities include the coding, recognition, interpretation, translation, and generation of language. [...] Advances in human language technology offer the promise of nearly universal access to on-line information and services. Since almost everyone speaks and understands a language, the development of spoken language systems will allow the average person to interact with computers without special skills or training, using common devices such as the telephone. These systems will combine spoken language understanding and generation to allow people to interact with computers using speech to obtain information on virtually any topic, to conduct business and to communicate with each other more effectively. (Cole 1996:1)

It is the aim of this study to explore some of the aspects and challenges in Human Language Technologies that are of relevance to computer-assisted language learning. Starting with a brief outline of some of the early attempts in HLT, namely in machine translation, it will become apparent that experiences and results in this area had a direct bearing on some of the developments in CALL. CALL soon became a multi-disciplinary field of research, development and practice. Some researchers began to develop CALL applications that made use of Human Language Technologies, and a few such applications will be brought to attention in this study. The advantages and limitations of applying HLT to CALL will be discussed, using the example of parser-based CALL. This brief discussion will form the basis for first hypotheses about the nature of human-computer interaction (HCI) in parser-based CALL.

2. Computers and language: A short look back

Facilitating and supporting all aspects of human communication through machines has interested researchers for a number of centuries. The use of mechanical devices to overcome language barriers was proposed first in the seventeenth century. Then, suggestions for numerical codes to be used to mediate between languages were made by Leibnitz, Descartes and others (cf. Hutchins 1986:21). The beginnings of what we describe today as Human Language Technologies are, of course, closely connected to the advent of computers:

> The electronic digital computer was a creation of the Second World War: the ENIAC machine at the Moore School of Electrical Engineering in the University of Pennsylvania was built to calculate ballistic firing tables; the Colussus machine at Bletchley Park in England was built to decipher German military communications. (Hutchins 1986:24)

In a report written in 1948, Turing, one of the fathers of Artificial Intelligence, who was heavily involved in the cryptoanalysis using the Colossus machine at Bletchley Park, mentions a number of different ways in which these new computers could demonstrate their "intelligence":

> "(i) Various games, e.g. chess, noughts and crosses, bridge, poker; (ii) *The learning of languages*; (iii) translation of languages; (iv) Cryptography; (v) mathematics (Turing 1948[2])". (Hutchins 1986:26f.; italics added)

The term "machine translation" was coined shortly after the first computers and computer programs had been produced in March 1947 by Booth and Weaver. Machine translation enjoyed a period of popularity with researchers and funding bodies in the United States and the Soviet Union:

> From 1956 onwards, the dollars (and roubles) really started to flow. Between1956 and 1959, no less than twelve research groups became established at various US universities and private corporations and research centres. [...] The kind of optimism and enthusiasm with which researchers tackled the task of MT [machine translation] may be illustrated best by some prophecies of Reifler, whose views may be taken as representative of those of most MT workers at that time: '... it will not be very long before the remaining linguistic problems in machine translation will be solved for a number of important languages' (Reifler 1958:518), and '... in about two years (from August 1957), we shall have a device which will at a glance read a whole page and feed what it has read into a tape recorder and thus remove all human co-operation on the input side of the translation machines' (Reifler 1958:516).[3] (Buchmann 1987:14)

Although linguists, language teachers and computer users today may find these predictions ridiculous, it was the enthusiasm and the work during this time that form the basis of many developments in HLT today.

By 1964, however, the promise of operational MT systems still seemed
distant and the sponsors set up a committee, which recommended in 1966
that funding for MT should be reduced. It brought to an end a decade of
intensive MT research activity. (Hutchins 1986:39)

It is then perhaps not surprising that the mid-sixties saw the birth of another discipline – Computer-Assisted Language Learning.[4] The *PLATO* Project is widely regarded as the beginning of CALL.[5] *PLATO IV* was probably the version of this
project (on mainframe computers) that had the biggest impact on the development
of CALL.[6] At the same time, another American university, Brigham Young University, received government funding for a CALL project, *TICCIT*, Time-Shared, Interactive, Computer Controlled Information Television (Levy 1997:18). Other still
well-known and widely used programs[7], for instance for German, were developed
soon afterwards: *CALIS,* Computer Aided Language Instruction System, at Duke
University (Borchardt:1995) and *TUCO* at Ohio State University. In Britain, John
Higgins developed *Storyboard* in the early 1980s, a total text-reconstruction program for the microcomputer.[8] The idea quickly caught on:

> Other programs such as *Fun with Texts* [Camsoft] extended the total text
> reconstruction idea considerably by adding further activities. (Levy 1997:25)

In recent years, the development of CALL has been greatly influenced by the technology and by our knowledge of and our expertise with it, so that not only the
design of most CALL software, but its classification has been technology-driven. A
classification which takes into account the most recent developments in Information Technology and European standards has been provided by Wolff (1993:21),
who distinguishes five groups of applications:

- Traditional computer-assisted language learning applications
- Artificial intelligence applications
- Utility applications (tools and help systems)
- Multi-media applications
- Communication applications

It has to be mentioned that most CALL packages combine features which make it
possible to assign them to more than one group. Traditional computer-assisted language learning applications are pieces of software which are designed very much
along the lines of the programmes which were produced in the early CALL projects.
They usually incorporate a tutorial element which introduces the concepts to be
learned and provides some help with the task at hand. This tutorial section is then
followed up by a set of exercises. The interface of these packages has improved
significantly due to the ability of computers to handle multimedia (text, sound,
picture, animation, video). Most of them offer a limited choice of exercise types

such as gap filling, text reconstruction, multiple choice, ranking, and provide feedback on the basis of more or less sophisticated pattern matching. These packages can be created using dedicated authoring tools (e.g. *WinCALIS, Question Mark Designer*), authoring programmes (e.g. *Guide, Toolbook, Authorware*) or, of course, high-level programming languages (e.g. *Delphi, Visual Basic*). Utility applications include special software like monolingual, bilingual and multilingual dictionaries as well as spell checkers; and software that has not been written with the language learner in mind such as word processors, databases and in some contexts even spreadsheets. Communication applications have become more popular over recent years because an increasing number of people have got access to the Internet. Electronic mail, bulletin boards, mailing lists, news groups, chat groups and computer conferencing, Web conferencing, video conferencing are examples of this technology that provide numerous application possibilities in language learning and teaching. The second half of the 1990s, in particular, saw an increased interest by language teachers in the use of the Internet and its most famous part, the World Wide Web (WWW). Web-based technology is used more and more, not just as a means of disseminating information for language learners, but also to provide easy access to (simple) exercises. Authoring tools for these exercises are now on the market (e.g. *Question Mark Perception, Hot Potatoes*).

The late 1980s saw the beginning of attempts which are mostly subsumed under *Intelligent CALL* (ICALL), a "mix of AI [Artificial Intelligence] techniques and CALL" (Matthews 1992b:i). Bowerman (1993) notes that:

> Weischedel et al. (1978) produced the first ICALL [Intelligent CALL] system which dealt with comprehension exercises. It made use of syntactic and semantic knowledge to check students' answers to comprehension questions. (Bowerman 1993:31)

As far as could be ascertained, this was just the early swallow that did not create a summer. Krüger-Thielmann (1992:51ff) lists and summarises the following early projects in ICALL: ALICE, ATHENA, BOUWSTEEN & COGO, EPISTLE, ET, LINGER, Menzel, Schwind, VP^2, XTRA-TE, Zock.[9] Matthews (1993:5) identifies Linguistic Theory and Second Language Acquisition Theory as the two main disciplines which inform Intelligent CALL and are (or will be) informed by Intelligent CALL and adds:

> [t]he obvious AI research areas from which ICALL should be able to draw the most insights are Natural Language Processing (NLP) and Intelligent Tutoring Systems (ITS). (Matthews 1993:6)

Matthews (1993:6) shows that it is possible to "conceive of an ICALL system in terms of the classical ITS architecture". The system consists of three modules – expert, student and teacher module – and an interface. The expert module is the one that "houses" the language knowledge of the system. It is this part which can process any piece of text produced by a learner – in an ideal system. This is usually done with the help of a parser of some kind:

> The use of parsers in CALL is commonly referred to as intelligent CALL or 'ICALL'; it might be more accurately described as *parser-based CALL* (italics added), because its 'intelligence' lies in the use of parsing – a technique that enables the computer to encode complex grammatical knowledge such as humans use to assemble sentences, recognise errors, and make corrections. (Holland et al. 1993:28)

This notion of parser-based CALL not only captures the nature of the field much better than the somewhat misleading term "Intelligent CALL" (Is all other CALL un-intelligent?), it also identifies the use of Human Language Technologies as one possible approach in CALL alongside others such as multimedia-based CALL and Web-based CALL and thus identifies parser-based CALL as one possible way forward for CALL. In some cases, the (technology-defined) borders between these sub-fields of CALL are not even clearly identifiable, as we will see in some of the projects mentioned later.

Of course, the previous sections are not intended to be an overview either of computer-assisted language learning or of parser-based CALL; these can be found elsewhere (e.g. *Computergestützter Fremdsprachenunterricht* 1985; Davies 1988; Levy 1997:13ff). Their main purpose here is to briefly illustrate that the link between technology in general and Human Language Technologies in particular and computer-assisted language learning is certainly not a recent development and draws on research and developments over a number of decades. However, even on the basis of these few remarks on CALL, its achievements and its history, and the many theoretical paradigms this area attempts to explore, apply and develop, we can draw the conclusion that one of the main features of CALL is its immense diversity in approaches as well as in underlying theories. It is this diversity which makes investigations in CALL so exciting and fruitful. On the other hand, the eclecticism towards neighbouring disciplines bears some dangers: important facts might have been ignored, or perhaps only misleading parts of a theory in a neighbouring discipline might have been borrowed, to the peril of CALL. Consequently, the set of theories that inform a particular research project in CALL will have to be limited in order to be fully explored and adequately applied.

3. HLT and learning: Some recent examples

The lack of linguistic modelling and the insufficient deployment of Natural Language Processing techniques has sometimes been given as one reason for the lack of progress in some areas of CALL (see, for example, Levy 1997:3, who quotes Kohn 1994:32[10]). Admittedly, there are few CALL applications which incorporate Natural Language Processing (NLP). In my opinion, this is due to the fact that parsing techniques and our knowledge about linguistics (cf. Matthews 1993 for an overview of grammatical theories applied in CALL) have only fairly recently reached a level which permits the development of fully functional parsers. In addition, the limited use of Human Language Technologies in CALL is due to the fact that the development of a parser, a computational grammar and their integration in a CALL package is a very complicated and time-consuming endeavour.

It will be shown in this section that it is possible to apply certain linguistic theories (e.g. phonology and morphology) to Human Language Technologies and implement this technology in CALL software. However, opponents of a parser-based approach in CALL claim that:

> AI architecture is still a long way from being able to create anything close to mirror the complex system of communication instantiated in any human language and is, hence, unable to introduce any qualitative leap in the design of CALL programs. (Salaberry 1996:11)

Salaberry backs up this claim by adding that:

> [t]he most important reason for this failure [of ICALL] is that NLP (Natural Language Processing) programs – which underlie the development of ICALL – cannot account for the full complexity of natural human languages. (Salaberry 1996:12)

This is, of course, true. However, it does not mean that interesting fragments or aspects of a given language cannot be captured by a formal linguistic theory[11] and hence implemented in a CALL application. This point, in defence of HLT and CALL, has been successfully argued by Nerbonne et al. (1998) in their introduction to the proceedings of the Language Teaching and Language Technology conference in Groningen in 1997 (Jager et al. 1998). At this conference, a number of successful implementations of HLT in CALL were presented. Carson-Berndsen demonstrated that:

> ... finite-state phonology has shown that the argument [by Salaberry (1996)] cannot be upheld for the phonological domain. The phonological knowledge base used [...] here is a complete and fully evaluated phonotactics of German. (Carson-Berndsen 1998:12)

APron, Autosegmental Pronunciation Teaching, uses this knowledge base and generates "event structures for some utterance" (Carson-Berndsen 1998:15) and can, for example, visualise pronunciation processes of individual sounds and well-formed utterances using an animated, schematic vocal tract. Witt & Young (1998), on the other hand, are concerned with assessing pronunciation. They implemented and tested a pronunciation scoring algorithm which is based on speech recognition and uses hidden Markov models:

The results show that – at least for this setup with artificially generated pronunciation errors – the GOP [goodness of pronunciation] scoring method is a viable assessment tool. (Witt & Young 1998:31)

A third paper on pronunciation, by Skrelin & Volskaja (1998) outlined the use of speech synthesis in language learning and lists dictation, distinction of homographs, a sound dictionary and pronunciation drills as possible applications. A number of papers, presented at this conference, are based on results of the *GLOSSER* project, a COPERNICUS project that aims to demonstrate the use of language processing tools (*Locolex*, a morphological analyser and part-of-speech disambiguation package from Rank Xerox Research Centre, relevant electronic dictionaries, such as *Hedendaag Frans*, and access to bilingual corpora):

The project vision foresees two main areas where *GLOSSER* applications can be used. First, in language learning and second, as a tool for users that have a bit of knowledge of a foreign language, but cannot read it easily or reliably. (Dokter & Nerbonne 1998:88)

Dokter & Nerbonne report on the French-Dutch demonstrator running on UNIX. The demonstrator:

- uses morphological analysis to provide additional grammatical information on individual words and to simplify dictionary look-up;
- relies on automatic word selection;
- offers the opportunity to insert glosses (taken from the dictionary look-up) into the text;
- relies on string-based word sense disambiguation: "Whenever a lexical context is found in the text that is also provided in the dictionary, the example in the dictionary is highlighted. (Dokter & Nerbonne 1998:93)

Roosma & Prószéky (1998) draw attention to the fact that *GLOSSER* works with the following language pairs: English-Estonian-Hungarian, English-Bulgarian, French-Dutch and describe a demonstrator version running under Windows. Dokter, Nerbonne, Schürcks-Grozeva & Smit conclude in their user study "that Glosser-RuG improves the ease with which language students can approach a foreign language text" (1998:175).

Other CALL projects that explore the use of language processing technology are RECALL[12] (Murphy et al. 1998; Hamilton 1998), a "knowledge-based error correction application" (Murphy et al. 1998:62) for English and German, and the development of tools for learning Basque as a foreign language (Diaz de Ilarranza et al. 1998). The latter project relies on a spell-checker, morphological analyser, syntactic parser and a lexical database for Basque, and the authors report on the development of an interlanguage model.

At another conference which brought together a group of researchers who are exploring the use of HLT in CALL software (Schulze et al. 1999), Tschichold (1999) discussed strategies for improving the success rate of grammar checkers. Menzel and Schröder (1999) described error diagnosis in a multi-level representation. The demonstration system captures the relations of entities in a simple town scenery. The available syntactic, semantic and pragmatic information is checked simultaneously for constraint violations, i.e. errors made by the language learners. Visser (1999) introduced *CALLex*, a program for learning vocabulary based on lexical functions. Diaz de Ilarraza et al. (1999) described aspects of *IDAZKIDE*, a learning environment for Spanish learners of Basque. The program contains the following modules: wide-coverage linguistic tools (lexical database with 65,000 entries; spell checker; a word form proposer and a morphological analyser), an adaptive user interface and a student modelling system. The model of the students' language knowledge, i.e. their interlanguage, is based on a corpus analysis (300 texts produced by learners of Basque). Foucou & Kübler (1999) presented a Web-based environment for teaching technical English to students of Computing. Ward et al. (1999) showed that Natural Language Processing techniques combined with a graphical interface can be used to produce meaningful language games. Davies & Poesio (1998) reported on tests of simple CALL prototypes that have been created using CSLUrp, a graphical authoring system for the creation of spoken dialogue systems. They argue that since it is evident that today's dialogue systems are useable in CALL software, it is now possible and necessary to study the integration of corrective feedback in these systems. Mitkov (1998) outlined plans for a new CALL project, The Language Learner's Workbench. It is the aim of this project to incorporate a number of already available HLT tools and to package them for language learners.

To conclude this section, three projects carried out at UMIST will briefly be mentioned. Bowerman (1993) developed a prototype, LICE, a program designed to support learners of German in writing texts. He provides a good overview of the history of parser-based CALL and links the projects he describes within this history with the development of Intelligent Tutoring Systems (ITS). Brocklebank (1998) worked with a parser developed by Ramsay (1999). Together with Ramsay he explored the implementation of relaxed constraints in an existing parser grammar for English. His study of the prototype implementation clearly shows that this approach to robust parsing of erroneous texts produced by language learners (in his case Japanese learners of English) is a very effective one. The same parser and grammar (Ramsay 1999) has been adapted for a number of languages including German (Ramsay & Schäler 1997; Ramsay & Schulze 2000). This parser forms the backbone of the *Textana* project which is a grammar checker for English learners of German (Schulze 1997; 1998). *Textana*'s approach is similar to Brocklebank's approach in that it relies on relaxed constraints (Schulze 1998), i.e. the parser does not fail when it encounters a violation of a grammatical rule, but instead records the nature of the violation and continues the parsing of the entire sentence.

These recent examples of CALL applications that make use of Human Language Technologies are by no means exhaustive. They not only illustrate that research in HLT in CALL is vibrant, but also that HLT has an important contribution to make in the further development of CALL. In the next section, the nature of this contribution will be discussed using the example of natural language parsers in CALL.

4. Parsers and language learning: A general discussion

Natural language parsers take written language as their input and produce a formal representation of the syntactic and sometimes semantic structure of this input.[13] The role they have to play in computer-assisted language learning has been under scrutiny in the last decade (Matthews 1992a; Holland et al. 1993; Nagata 1996). Holland et al. (1993:28) discussed the "possibilities and limitations of parser-based language tutors". Comparing parser-based CALL to what they label as conventional CALL they come to the conclusion that:

> ... in parser-based CALL the student has relatively free rein and can write a potentially huge variety of sentences. ICALL thus permits practice of production skills, which require recalling and constructing, not just recognising [as in conventional CALL], words and structures. (Holland et al. 1993:31)

However, at the same time, parsing imposes certain limitations. Parsers tend to concentrate on the syntax of the textual input, thus:

> ICALL may actually subvert a principal goal of language pedagogy, that of communicating meanings rather than producing the right forms. (Holland et al. 1993:32)

This disadvantage can be avoided by a "focus on form" which is mainly achieved by putting the parser/grammar checker to use within a relevant, authentic communicative task and at a time chosen by and convenient to the learner/text producer.

Juozulynas evaluated the potential usefulness of syntactic parsers in error diagnosis. He analysed errors in an approximately 400 page corpus of German essays by American college students in second-year language courses. His study shows that:

> ... syntax is the most problematic area, followed by morphology. These two categories make up 53% of errors [...] The study, a contribution to the error analysis element of a syntactic parser of German, indicates that most student errors (80%) are not of semantic origin, and therefore, *are potentially recognizable by a syntactic parser.* (Juozulynas 1994:5; italics added)

Juozulynas adapted a taxonomic schema by Hendrickson[14] which comprises four categories: syntax, morphology, orthography, lexicon. Juozulynas's argument for splitting orthography into spelling and punctuation is easily justified in the context of syntactic parsing. Parts of punctuation can be described by using syntactic bracketing rules, and punctuation errors can consequently be dealt with by a syntactic parser. Lexical and spelling errors form, according to Juozulynas, a rather small part of the overall number of learner errors. Some of these errors will, of course, be identified during dictionary look-up, but if words that are in the dictionary are used in a nonsensical way, the parser will not recognise them unless specific error rules (e.g. for false friends) are built in. It can be assumed, however, that the parser is used in conjunction with a spell-checker (so most of the orthographic errors could be eliminated before the parser starts its analysis) and that learners have good knowledge of what they want to say (to avoid many errors of a semantic nature). Consequently, a parser-based CALL application can play a useful role in detecting many of the morpho-syntactic errors which constitute a high percentage of learner errors in freely produced texts.

Nevertheless, the fact remains:

> A second limitation of ICALL is that parsers are not foolproof. Because no parser today can accurately analyse all the syntax of a language, false acceptance and false alarms are inevitable. (Holland et al. 1993:33)

This is something not only developers of parser-based CALL, but also language learners using such software have to take into account. In other words, this limitation of parser-based CALL has to be taken into consideration during the design and implementation process and when integrating this kind of CALL software in the learning process.[15]

> A final limitation of ICALL is the cost of developing NLP systems. By comparison with simple CALL, NLP development depends on computational linguists and advanced programmers as well as on extensive resources for building and testing grammars. Beyond this, instructional shells and lessons must be built around NLP, incurring the same expense as developing shells and lessons for CALL. (Holland et al. 1993:33)

It is mainly this disadvantage of parser-based CALL that explains the lack of commercially available (and commercially viable) CALL applications which make good use of HLT. However, it is to be hoped that this hurdle can be overcome in the not too distant future because sufficient expertise in the area has accumulated over recent years. More and more computer programs make good use of this technology, and many of these have already "entered" the realm of computer-assisted language learning, as can be seen from the examples in the previous section.

Holland et al. (1993) answer their title question "What are parsers good for?" on the basis of their own experience with *BRIDGE*, a parser-based CALL program for American military personnel learning German, and on the basis of some initial testing with a small group of experienced learners of German. They present the following points:

- *ICALL appears to be good for form-focused instruction* offering learners the chance to work on their own linguistic errors by this method, not only to improve their performance in the foreign language but also to improve their language awareness
- *ICALL appears to be good for selected kinds of students.* The authors list the following characteristics which might influence the degree of usefulness of ICALL for certain students:
  - intermediate proficiency
  - analytical orientation
  - tolerance of ambiguity
  - confidence as learners
- *ICALL is good for research* because the parser automatically tracks whole sentence responses and detects, classifies, and records errors. It might facilitate the assessment of the students grammatical competency and thus help us discover patterns of acquisition
- *ICALL [...] can play a role in communicative practice.* The authors argue for the embedding of parser-based CALL in "graphics microworlds" which help to capture some basic semantics.

Given that we would like to harness the advantages of parser-based CALL, how do we take the limitations into consideration in the process of designing and implementing a parser-based CALL system? What implications does the use of parsing technology have for human-computer interaction?

5. Computers and learning: Some implications

> This limitation [that parsers are not foolproof] leads to a second disadvantage of ICALL: Because parsers give the illusion of certainty, their feedback may at best confuse students. [...] By contrast, in traditional CALL the very rigidity of the response requirement guarantees certainty. [...] [I]n discourse analytic terms (Grice 1975), the nature of the contract between student and CALL tutor is straightforward, respecting the traditional assumption that the teacher is right, whereas the ICALL contract is less well defined. (Holland et al. 1993:33f)

The rigidity of traditional CALL in which the program controls the linguistic input by the learner to a very large extent has often given rise to a criticism of CALL which accuses CALL developers and practitioners of relying on behaviourist programmed instruction.[16] If one wants to give learners full control over their linguistic input, for example, by relying on parsing technology, what are then the terms according to which the communicative interaction of computer (program) and learner can be defined? The differences between humans and machines have obviously to be taken into consideration in order to understand the interaction of learners with CALL programs:

- Machines are compiled out of individual parts for a very specific purpose (*totum fix et partibus*); whereas humans are holistic entities whose parts can be differentiated (*partes fiunt ex toto*).
- Humans process all sorts of experiences and repeatedly and interactively create their own environment – something machines cannot do. They "calculate" a problem on the basis of pre-wired rules.
- The main features of human thoughts are the inherent contradictions and the ability to cope with them, something that will not be calculable due to its complexity, variety and degree of detail (Schmitz 1992:209f; my translation).

Earlier Schmitz argues that:

... our world of thought is split into two realms: the abstract example – the lacking reality, the perfect plan – the resistance against its realisation, the universal rules – the numerous exceptions, the manageable model – the real chaos, the "pure" ideal – the "dirty" reality [...] and the technically masterable – the rebellious everyday life. This approach was thoroughly rejected by Ludwig Wittgenstein. He radically overturned the thinking in ideas, abstractions and identities, showing in the same line of argument that rules do not exist in independence of their application and that we change them by using them. Beyond their application we do not have any control over them. [...] Because they [computers] are instruments, language can only be used in an instrumental way with them. Tools embody abstract technical knowledge; language-enabled computers are abstract incarnations of human knowledge about rules. The computer awaits its application without being able to significantly change within this application. (Schmitz 1992:62; my translation)

These differences between humans and machines can for our purposes, i.e. the theoretical description of human-computer interaction, be legitimately reduced to the distinction between actions and operations as is done in Activity Theory. Zech[17] lists four major outcomes of this theory:

- the knowledge that linguistic-communicative activity is embedded in the more general context of human activity and is motivated by this context; as a result the requirement in teaching methodology to organise linguistic learning on the basis of motivated linguistic activities receives renewed confirmation and justification;

- the knowledge that the production and reception of linguistic utterances can be modelled as process in which certain stages, phases and sub-processes of cognitive-linguistic activity can be distinguished, this makes the intentionality and systematicity of the pedagogic-methodological impact more transparent;
- the knowledge that linguistic-communicative activity is realised in individual intentional actions and that the carrying out of such actions is based on the carrying out of sub-actions (operations); from this one can conclude for teaching methodology that the intentional training of complex qualities of communicative actions has to consider the sub-actions necessary for those actions;
- the knowledge that actions and sub-actions can be converted in either direction, i.e. sub-actions can be raised to the level of intentional actions, actions can be integrated as sub-actions (operations) into other actions; from the point of view of teaching methodology this is of importance for the step-by-step organisation of the ability to communicate, especially for the systematic training of sub-actions at the level of conscious actions and for their transferral into an integrated application (cf. the example of changing gear in the process of learning to drive given by A. N. L'eont'ev). (Zech 1985:193)

The main point here is that communicative activities can be divided into actions which are intentional, i.e. goal-driven; and these can be sub-divided into operations which are condition-triggered. These operations are normally learnt as actions. In the example referred to in the quotation above, the gear-switching is learnt as an action. The learner-driver is asked by the driving instructor to change gear and this becomes the goal of the learner. Once the learner-driver has performed this action a sufficient number of times, this action becomes more and more automated and in the process loses more and more of its intentionality. A proficient driver might have the goal to accelerate the car which will necessitate switching into higher gear, but this is now triggered by a condition (the difference between engine speed and speed of the car). It can thus be argued that humans learn to perform complex actions by learning to perform certain operations in a certain order. Machines, on the other hand, are only made to perform certain (sometimes rather complex) operations. Sequences of such operations can be performed by sophisticated machines like computers in rapid succession (or with powerful computers even in parallel), but each and every one of them is initialised by a condition (e.g. a mouse-click, keyboard input) and they are not subordinated to an intention, so they cannot be described as actions (cf. Schmitz 1992:169).

  This has some bearing on our understanding of the human-computer interaction that takes place when a learner uses language learning software. When, for instance, the spell checker is started in a word-processing package, the software certainly does not have the intention to proof-read the learner's document. The computer just responds to the clicking of the spellchecker menu item and performs the operation of checking the strings in the document against the entries in a machine dictionary. The result of the comparison triggers the next operation: if an identical string has been found in the dictionary, the next string from the document is being compared; if an identical string could not be found in the dictionary, similar strings will be selected from the dictionary according to an established algorithm and these will be displayed to the user as spelling alternatives. For the computer user (in our case a learner), it might look like the computer is proof-reading the document. Normally, one only realises that no "proper" document checking is going on if a correctly spelled word is not found in the dictionary or nonsense alternatives are given for a simple spelling error.

Person X interacts with Person Y in that he observes Person Y's action, reasons about the likely intention for that action and reacts according to this assumed intention. It appears to be the case that many learners transfer this approach to the interaction with a computer in a language learning situation, i.e. they interpret the sequence of operations performed by the computer as an action, reason about the "intention" of the computer and react accordingly. This, for example, explains why many learners, when an answer they believe to be right is rejected by the computer, get just as frustrated as they would get if it were rejected by their tutor.

Of course, an ideal computer-assisted language learning system would avoid such pitfalls and not reject a correct response or overlook an incorrect one. Since any existing system can only approximate to this ideal, researchers and developers in parser-based CALL can only attempt to build systems that can perform complex structured sequences of (linguistic) operations so that learners can interact meaningfully and successfully with the computer.

Many of the CALL programs and prototypes in parser-based CALL and related areas show that HLT is already contributing to further advances in computer-assisted language learning by giving the learner more control over the communicative interaction and by offering more flexible support for tutors. Perhaps it is at least as important that research in HLT – in computational and formal linguistics, in artificial intelligence; in natural language processing and machine translation – can greatly contribute to further advances in research in computer-assisted language learning.

Notes

[1] The term "Human Language Technologies" (HLT) was officially introduced by the European Commission in January 1999 to replace the term "Language Engineering". HLT also embraces the area which is known as "Natural Language Processing": see Jager's article in this publication. See also the HLT website at: http://www/linklink/lu.

[2] Turing, A. M. (1948) Intelligent machinery. In B. Meltzer & D. Mitchie, (Eds) (1969) *Machine intelligence 5* (pp. 3–23). Edinburgh: University Press.

[3] Reifler, E. (1958) The Machine Translation Project at the University of Washington, Seattle. In *Proceedings of the International Congress of Linguists, Oslo* (pp. 514–518). Washington: University of Washington.

[4] "In order to explain how CALL originated and the form its development took, it is necessary first briefly to consider the history of the application of the computer to language and literature in general. And in the context of cryptographical investigations

in the Second World War, language applications were involved in the very earliest developments of the electronic digital computer" (Last 1992:227).

5   "CALL may be said to have begun with the PLATO (Programmed Logic for Automatic Teaching Operations) Project which was initiated at the University of Illinois in 1960" (Levy 1997:5).

6   "Over the years, we explored many CALL design problems such as the use of hierarchical menus, the organization of student controlled help systems, the proper use of automatic review features, and the possibilities of matching-driven grammar error diagnosis, feedback, and error markup. We experimented extensively with multilingual writing systems, touch driven selection activities, combination of graphics and audio with text material, and performance history and data basing. The solutions to these problems eventually provided a rich design vocabulary for PLATO CALL. Indeed, before the appearance of true multimedia courseware, I never saw a micro based CALL design we hadn't already approximated on PLATO" (Hart 1995:25).

7   See also Sanders (1995). Introduction to *Thirty years of Computer-Assisted Language Instruction*, Special Issue of *CALICO Journal, 12.4*, recalling the beginnings of CALL with the Stony Brook Project, PLATO, TICCIT and CALIS in the United States.

8   Davies (1996:1) notes that the original idea of text-reconstruction programs for language learning, which gave birth to programs such as *Storyboard*, probably emanates from Tim Johns.

9   The words in capitals refer to the project title, whereas the names give the main author of a paper on one of the projects if no project title was given.

10   Kohn, K. (1994). Distributive language learning in a computer-based multilingual communication environment. In H. Jung & Vanderplank, R. (Eds), *Barriers and bridges: media technology and language learning: proceedings of the 1993 CETall Symposium on the occasion of the 10th AILA World Congress in Amsterdam*. Frankfurt: Peter Lang.

11   For the place and nature of a formal linguistic description in CALL, see Schulze (1999).

12   RECALL: Repairing Errors in Computer Aided Language, EU Telematics Applications of Common Interest – Language Engineering LE1-1615.

13   For an introduction see, for example, Allen (1995).

14   Hendrickson, J.M. (1979). Evaluating spontaneous communication through systematic error analysis. In *Foreign Language Annals, 12.5*, 357–364.

15   See the discussion of these implications in the section "Computers and learning: some Implications".

16   For an overview of behaviourist learning theories, see, for example Curzon (1985:6ff).

17   Zech appears to base his discussion mainly on A. N. L'eont'ev, but there are at least two more names inextricably linked with Activity Theory, the psycholinguist A. A. L'eont'ev and, of course, L. S. Vygotsky.

References

– (1985). *Computergestützter Fremdsprachenunterricht. Ein Handbuch.* (Herausgegeben von der Langenscheidt-Redaktion). Berlin: Langenscheidt.

– (1996). *Language and technology: From the Tower of Babel to the Global Village.* Brussels: European Commission.

Allen, J. (1995). *Natural language understanding*. New York: Benjamins/Cummings Publishing Company.

Borchardt, F. (1995). Language and computing at Duke University: Or, Virtue Triumphant, for the time being. *CALICO Journal, 12.4*, 57–83.

Bowerman, C. (1993). *Intelligent computer-aided language learning. LICE: A system to support undergraduates writing in German*. Unpublished doctoral dissertation, Manchester, UMIST.

Brocklebank, P. (1998). *An experiment in developing a prototype intelligent teaching system from a parser written in Prolog*. Unpublished MPhil dissertation, Manchester, UMIST.

Buchmann, B (1987). Early history of Machine Translation. In M. King (Ed.), *Machine Translation today: The state of the art* (pp. 3–21). Edinburgh: University Press.

Carson-Berndson, J. (1998). Computational autosegmental phonology in pronunciation teaching. In S. Jager, Nerbonne J. & van Essen A.(Eds), *Language teaching and language technology* (pp. 11–20). Lisse: Swets & Zeitlinger.

Cole, R. (1996). Foreword. In R. Cole (Ed. in-chief), *Survey of the state of the art in Human Language Technologies*. Available (2000): http://cslu.cse.ogi.edu/HLTsurvey/HLTsurvey.html

Curzon, L.B. (1985). *Teaching in further education: An outline of principles and practice* (3rd edition). London: Holt, Rinehart and Winston.

Davies, G. (1988). CALL software development. In U. Jung (Ed.), *Computers in applied linguistics and language learning: A CALL handbook* (pp. 29–47). Frankfurt: Peter Lang.

Davies, G. (1996). *Total-text reconstruction programs: A brief history*. Maidenhead: Camsoft Monograph (unpublished manuscript).

Davies, S. & Poesio, M. (1998, May). *The provision of corrective feedback in a spoken dialogue system*. Paper presented at the conference on NLP in CALL, Manchester, UMIST.

Diaz de Ilarranza, A., Maritxalar, M. & Oronoz, M. (1998). Reusability of language technology in support of corpus studies in an ICALL environment. In S. Jager, Nerbonne J. & van Essen A. (Eds), *Language teaching and language technology* (pp. 149–166). Lisse: Swets & Zeitlinger.

Diaz de Ilarranza, A., Maritxalar, A., Maritxalar, M. & Oronoz, M. (1999). IDAZKIDE: An intelligent computer-assisted language learning environment for Second Language Acquisition. In M. Schulze, Hamel M-J. & Thompson J.

(Eds), *Language processing in CALL, ReCALL* Special Edition (pp.12–19). Hull: CTI Centre for Modern Languages, University of Hull.

Dokter, D. & Nerbonne, J. (1998). A session with Glosser-RuG. In S. Jager, Nerbonne J. & van Essen A. (Eds), *Language teaching and language technology* (pp. 89–95). Lisse: Swets & Zeitlinger.

Dokter, D., Nerbonne, J., Schurcks-Grozeva, L. & Smit, P. (1998). Glosser-RuG: A user study. In S. Jager, Nerbonne J. & van Essen A. (Eds), *Language teaching and language technology* (pp. 169–178). Lisse: Swets & Zeitlinger.

Foucou, P.-Y. & Kübler, N. (1999). A Web-based language learning environment: General architecture. In M. Schulze, Hamel M-J. & Thompson J. (Eds), *Language processing in CALL, ReCALL* Special Edition (pp. 31–39). Hull: CTI Centre for Modern Languages, University of Hull.

Hamilton, S. (1998) A CALL user study. In S. Jager, Nerbonne J. & van Essen A. (Eds), *Language teaching and language technology* (pp. 200–208). Lisse: Swets & Zeitlinger.

Hart, R. (1995). The Illinois PLATO Foreign Languages Project. *CALICO Journal, 12.4*, 15–37.

Hendrickson, J.M. (1979). Evaluating spontaneous communication through systematic error analysis. In *Foreign Language Annals, 12.5*, 357–364.

Holland, M., Maisano, R., Alderks, C. & Martin, J. (1993). Parsers in tutors: What are they good for?. *CALICO Journal, 11.1*, 28–46.

Hutchins, J. (1986). *Machine Translation – past, present and future.* New York: Ellis Horwood Ltd.

Jager, S., Nerbonne, J. & van Essen, A. (Eds) (1998). *Language teaching and language technology.* Lisse: Swets & Zeitlinger.

Juozulynas, V. (1994). Errors in the composition of second-year German students: An empirical study of parser-based ICALI. *CALICO Journal, 12.1*, 5–17.

King, M. (Ed.) (1987). *Machine Translation today: The state of the art.* Edinburgh: University Press.

Kohn, K. (1994). Distributive language learning in a computer-based multilingual communication environment. In H. Jung & Vanderplank, R. (Eds), *Barriers and bridges: media technology and language learning: proceedings of the 1993 CETall Symposium on the occasion of the 10th AILA World Congress in Amsterdam.* Frankfurt: Peter Lang.

Krüger-Thielmann, K. (1992). *Wissensbasierte Sprachlernsysteme. Neue Möglichkeiten für den computergestützten Sprachunterricht.* Tübingen: Gunter Narr Verlag.

Last, R. (1992). Computers and language learning: Past, present – and future? In C. Butler (Ed.), *Computers and written texts* (pp. 227–246). Oxford: Blackwell.

Levy, M. (1997). *CALL: Context and conceptualisation.* Oxford: Oxford University Press.

Matthews, C. (1992a). Going AI. Foundations of ICALL. *CALL, 5.1–2,* 13–31.

Matthews, C. (1992b). *Intelligent CALL (ICALL) bibliography.* Hull: CTI Centre for Modern Languages, University of Hull.

Matthews, C. (1993) Grammar frameworks in Intelligent CALL. *CALICO Journal, 11.1,* 5–27.

Menzel, W. & Schröder, I. (1998). Error diagnosis for language learning systems. In M. Schulze, Hamel M-J. & Thompson J. (Eds), *Language processing in CALL, ReCALL* Special Edition (pp. 20–30). Hull: CTI Centre for Modern Languages, University of Hull.

Mitkov, R. (1998, May). *Language Learner's Workbench.* Paper presented at the conference on NLP in CALL, Manchester, UMIST. Available (2000): http://www.wlv.ac.uk/~le1825/projects/llw.html

Murphy, M., Krüger, A. & Griesz, A., (1998). RECALL – towards a knowledge-based approach to CALL. In S. Jager, Nerbonne J. & van Essen A. (Eds), *Language teaching and language technology* (pp. 62–73). Lisse: Swets & Zeitlinger.

Nagata, N. (1996). Computer vs. workbook instruction in Second Language Acquisition. *CALICO Journal, 14.1,* 53–75.

Nerbonne, J., Jager, S. & van Essen, A. (1998). Introduction. In S. Jager, Nerbonne J. & van Essen A. (Eds), *Language teaching and language technology* (pp. 1–10). Lisse: Swets & Zeitlinger.

Ramsay, A. (1999). *Demonstration of the parser.* Manchester: UMIST. Available (2000): http://www.ccl.umist.ac.uk/staff/allan

Ramsay, A. & Schäler, R. (1997). Case and word order in English and German. In R. Mitkov & N. Nicolov (Eds), *Recent advances in Natural Language Processing* (pp. 15–34). Amsterdam: John Benjamin.

Ramsay, A. & Schulze, M. (2000). Die Struktur deutscher Lexeme. *German Linguistic and Cultural Studies.* Frankfurt: Peter Lang.

Reifler, E. (1958) The Machine Translation Project at the University of Washington, Seattle. In *Proceedings of the International Congress of Linguists, Oslo* (pp. 514–518). Washington: University of Washington.Roosmaa, T. & Prószéky, G. (1998). GLOSSER – using language technology tools for reading texts in a foreign language. In S. Jager, Nerbonne J. & van Essen A. (Eds), *Language teaching and language technology* (pp. 101–107). Lisse: Swets & Zeitlinger.

Salaberry, R. (1996). A theoretical foundation for the development of pedagogical tasks in computer mediated communication. *CALICO Journal, 14.1*, 5-34.

Sanders, R. (1995). Thirty years of computer assisted language instruction: Introduction. *CALICO Journal, 12.4*, 5-14.

Schmitz, U. (1992). *Computerlinguistik*. Opladen: Westdeutscher Verlag.

Schulze, M. (1997). Textana – text production in a hypertext environment. *CALL, 10.1*, 71–82.

Schulze, M. (1998). Teaching grammar – learning grammar. Aspects of Second Language Acquisition in CALL. *CALL, 11.2*, 215–228.

Schulze, M. (1999). From the developer to the learner: Computing grammar – learning grammar. *ReCALL, 11.1*, 117–124.

Schulze, M., Hamel, M-J. & Thompson, J. (Eds) (1999). *Language processing in CALL, ReCALL* Special Edition. Hull: CTI Centre for Modern Languages, University of Hull.

Skrelin, P. & Volskaja, N. (1998). The application of new technologies in the development of education programs. In S. Jager, Nerbonne J. & van Essen A. (Eds), *Language teaching and language technology* (pp. 21–24). Lisse: Swets & Zeitlinger.

Tschichold, C. (1999). Intelligent grammar checking for CALL. In M. Schulze, Hamel M-J. & Thompson J. (Eds), *Language processing in CALL, ReCALL* Special Edition (pp. 5–11). Hull: CTI Centre for Modern Languages, University of Hull.

Turing, A. M. (1948) Intelligent machinery. In B. Meltzer & D. Mitchie, (Eds) (1969) *Machine intelligence 5* (pp. 3–23). Edinburgh: University Press.

Visser, H. (1999). CALLex (Computer-aided learning of lexical functions) – a CALL game to study lexical relationships based on a semantic database. In M. Schulze, Hamel M-J. & Thompson J. (Eds), *Language processing in CALL, ReCALL* Special Edition (pp. 50–56). Hull: CTI Centre for Modern Languages, University of Hull.

Ward, R., Foot, R. & Bostron, A.B. (1999). Language processing in computer-assisted language learning: Language with a purpose. In M. Schulze, Hamel M-J. & Thompson J. (Eds), *Language processing in CALL, ReCALL* Special Edition (pp. 40–49). Hull: CTI Centre for Modern Languages, University of Hull.

Weischedel, R., Vogel, W. & Jarvis, M. (1978). An Artificial Intelligence approach to language instruction. *Artificial Intelligence, 10*, 225–241.

Witt, S. & Young, S. (1998). Computer-assisted pronunciation teaching based on automatic speech recognition. In S. Jager, Nerbonne J. & van Essen A. (Eds),

Language teaching and language technology (pp. 25–35). Lisse: Swets & Zeitlinger.

Wolff, D. (1993). New technologies for foreign language teaching. In *Foreign language learning and the use of new technologies* (pp. 17–27). Brussels: Bureau Lingua.

Zech, J. (1985). Methodische Probleme einer tätigkeitsorientierten Ausbildung des sprachlich-kommunikativen Könnens. In G. Michel (Ed.), *Grundfragen der Kommunikationsbefähigung* (pp. 191–208). Leipzig: Bibliographisches Institut.

9

Learning out of control: Some thoughts on the World Wide Web in learning and teaching foreign languages

Thomas Vogel

Europa-Universität Viadrina, Frankfurt (Oder), DE

1. Introduction

For thousands of years, human beings have been learning foreign languages for various purposes, one way or the other, in different settings with more or less success. The World Wide Web is often claimed to have added new dimensions to this human activity. The aim of this chapter is to look critically at some of those dimensions and ask to what extent they can enrich language learning and teaching. This article is not based on extensive research relating to the Web as a language-learning environment, since an extensive body of theoretical and empirical research does not yet exist, but rather it attempts to ask questions which may motivate future research. The time is ripe for enthusiastic assumptions about the Web as a powerful tool in language learning and teaching to be backed up or disproved by empirical research on the Web as a learning environment.

The arguments on which this chapter is based originate from three areas of personal experience: The first is that of an English language teacher who experiments with the Web in his teaching without having found the final recipe for integration. The second is that of a language centre manager at a German university, responsible for providing resources and further training for students and staff. Here my interest is not restricted to English but extends to the eight European languages offered at the Centre (English, German, Finnish, French, Polish, Spanish, Swedish and Russian.). Thirdly, I am interested in the Web as a researcher who for a number of years has studied learners acquiring languages in different settings. My special focus in SLA-research has always been the so-called naturalistic settings, i.e. contexts in which learners acquire languages without extensive formal teaching (learning outside the language classroom). The question arises whether the Web can be said to constitute a new naturalistic environment.

Writing about the World Wide Web in language teaching is not without its dangers; it is rather like writing about "the book in language teaching". It is nevertheless important to address the general issues which the Web raises in language learning and to keep a broad general perspective, taking the risk of disappointing readers in search of "hot" websites and suggestions for concrete exciting Web projects. After an initial examination of approaches to the Web, the chapter will provide information on a small exploratory survey of students' use of the Web and their perception of its role in their language learning. This will be followed by an examination of the role of the Web as a context for naturalistic language acquisition and brief comments on Web pages specially designed for language learning. This will enable us to investigate the main methods of integrating the Web in the language classroom. Finally, some theoretical implications of using the Web in the language classroom will be outlined.

Research on the World Wide Web in language teaching and learning, usually conducted by enthusiastic practitioners in the field, is beginning to appear. It has hitherto focused on learner attitudes, on retention of factual information in cultural learning and on learner strategies (Osuna & Meskill 1998, Ganderton 1998, Warschauer 1997, Negretti 1999). The results clearly provide evidence for the motivational qualities of the Web. Research, however, has not yet focused on language learning results per se (see Warschauer & Healey 1998 for a similar view on the situation). It so far has been based on single, small-scale projects and was for the most part action research, largely driven by the question of whether the students actually enjoyed the project. There is ample evidence from SLA research that motivation is one of the factors in language learning, but it is definitely only one among many and it may not be as decisive as some teachers assume.

A research agenda, still to be developed, should focus on the Web as a new learning environment that combines features of classroom learning and of naturalistic settings. The research questions of whether learners actually progress to proficiency faster (rate of acquisition) and whether they take the same route to proficiency (route of acquisition) have not been addressed yet. This issue calls for longitudinal, comparative studies and long-term observation. In their overview on "Computers and Language Learning" Warschauer and Healey (1998) show that it will not be necessary to develop new research paradigms. They already exist in the fields of second language acquisition, psychology, FLT and presumably media studies. Those fields also provide sophisticated research tools. They should be brought together to form an interdisciplinary approach to Web-based language learning.

2. Approaches to the Web

With regard to the use of the Web for language teaching, two different approaches have evolved, the differences being in the didactic philosophy and in the method of Web integration. Both approaches are justifiable, as they definitely serve learner needs. The final products, however, i.e. the websites, are essentially different.

2.1 The medium-centred approach

The starting point of this approach is usually the fascination with the medium itself: "What is clear about the development is that pedagogy has been driven largely by technology [...] Developers have been excited from the start by what could be done with little effort..."(Felix 1999). Moreover, proponents of this approach justify the need to integrate the Web into language teaching and learning by the fact that the Web is becoming an integral part of both professional life and spare-time activities, at least in Western Europe and the US. Secondly, because the Web is becoming increasingly omnipresent in education in general, they try to find out how its potential can best be tapped for language learning and teaching. Human-machine interaction for self-study and elaborate design features are usually the characteristic features of this approach. Learners have to answer multiple choice questions for text comprehension, fill in gaps for grammar training, etc. They type in their answers, and the computer checks them and presents the results, depending on the technological sophistication of those self-testing mechanisms. The methodology behind medium-centred websites is usually well known from traditional textbooks.

2.2 The problem-solving approach

The second approach takes the shortcomings of traditional language teaching as a starting point. There is basically one key word that characterises the major problem of the traditional classroom: *authenticity*. In the classroom learners and teachers communicate in order to learn. The content of the communicative exchanges is more or less artificial, or at least it plays a secondary role. The problem-solving approach tries to make language learning more authentic. Its characteristic features are the use of real, as opposed to learning websites and the creation of environments for communicative human-to-human exchanges. The following quotation clearly summarises the main ideas of the problem-solving approach:

> When you combine student projects, Web publishing, and the communications power of the Internet, your teaching will be transformed. Your classroom walls will tumble down... Learning will become more authentic and more purposeful. (Global SchoolNet Foundation 1999)

Rüschoff and Wolff (1999) go as far as to claim that the Web actually dissolves the dichotomy between real life and life in the classroom.

It is often claimed that the authenticity of the medium leads to direct access to the target culture. Students are given the opportunity for virtual travelling without having to pay for the ticket. Thus, the Web is becoming an essential ingredient of intercultural communication, the gateway to foreign cultures. Students can read Web versions of daily newspapers and same-day news reports from sources such as the French Embassy's gopher service, the daily Revue de Press, etc. Such experiences can allow learners to participate in the culture of the target language, which in turn can enable them to further learn how cultural background influences one's view of the world (Singhal 1997).

Two further fundamental problems of teaching seem closer to a solution: motivation and learner autonomy. Looking at students surfing the Web for hours will
leave you without any doubt as far as the motivating power of the Web is concerned. The accessibility of the Web, the independence of space and time seems to
make it a perfect tool for individual, self-directed learning.

> When you look beneath the hype and the criticism of the Internet in schools,
> and observe students and teachers using the Internet in the ways we illus-
> trate here, you'll see them engaged, involved and active learners and teach-
> ers... both students and adults. (Global SchoolNet Foundation 1999)

All these claims raise a number of interesting issues and important questions. Firstly
what is meant by authenticity in the context of the Web? The debate on the definition of authenticity and authentic materials has been gaining in momentum since
the 1970s, and any detailed consideration of it is beyond the scope of this chapter.
From the original definition of an authentic text as one not created specifically for
the purpose of language learning, researchers such as Widdowson (1980), Van Lier
(1996) and many others have considered authenticity in relation not only to the text
but also in relation to the task and to the learner. The question then arises as to the
authenticity of Web-based material. The Web is part of the media and therefore has
media qualities. It presents one, maybe an extremely limited aspect of a culture.
Perhaps it is a culture or a subculture in its own right. For example, recent investigations into the language of electronic communication shows that it exhibits certain
characteristics that set it apart from non-electronic communication (see Collot &
Belmore 1996). Lastly although the Web itself has an addictive quality, especially
but not only for young people, it is not certain that learners can reconcile their
spare-time Web-surfing with their concept of learning. Will they accept guidance
on how they should use the Web for learning and can they be convinced that it will
be more effective than listening to cassettes and reading books, or at least just as
good?

3. The average surfer and language learner: A random check

In order to find out what university students do when surfing the Web, I asked 55
randomly selected students at the Europa-Universität Viadrina in Frankfurt (Oder),
Germany, who take foreign languages as a core element of their business or cultural
studies degree, how they make use of the resources of the Web. For some years the
Language Centre had made attempts to convince the students of the value of the
Web as a learning tool. Now I wanted to know whether they consciously use the
Web for language learning. All students have access to the Web, either at the university computer pools, the self-access centre or from the halls of residence. It
emerged that most of the students regularly surf the Web for various reasons. Their

main interest and focus is on their studies. Nearly all of the students visit websites in two (English, German) or more languages (French, Spanish, Polish, Swedish, Czech, Finnish and Italian). When asked whether they visit the Web to learn foreign languages, all of them claimed that this was definitely not their motivation. Interestingly, however, they somehow suspected that their Web activities, mainly searching for information and reading, foster unintentional learning. Below are some answers to the question on how they thought they were learning languages with the Web:

- "... I read pages in them ..."
- "... not consciously ..."
- "... just by reading ..."
- "... more, by the way ..."
- "... read and try to understand ..."
- "... maybe while reading ..."

When asked, none of the students could name a website for language learning. To conclude, these university students see the Web as an important tool to retrieve information within the context of their studies. They see the enormous learning potential. Nevertheless, for them language learning is considered to be rather a by-product than a conscious focus.

4. The Web: A context for real communication and naturalistic language acquisition

To what extent is a visit to the Web similar to a visit or a stay in a target-language environment? Does the Web exhibit the features of a naturalistic environment that offers more opportunities for unintentional learning?

The most noticeable similarity between the Web and real life has to do with the input with which learners are confronted. As soon as they enter the Web they are faced with an information overload on all communicative channels. After some initial shock, learners will realise that one of the basic skills is the ability to be selective, to concentrate on the pieces of information that they need or in which they are interested. Selectivity is usually not a skill which the learner needs in the classroom where the input has normally been selected by the teacher. Selectivity skills are also needed to evaluate the quality of information on the Web. Indeed quality and the evaluation of websites have become key issues recently (see Grassian 1997). However, this is not a Web-specific problem, for the ability to distinguish between useful and useless, true and false information, is a skill that human beings need in their daily life in order to survive. The quality of Web sources cannot or should not be evaluated in categories different from those which are used in evaluating books, newspapers, films, and the behaviour of human beings.

Other features of the Web are similar to those of naturalistic environments. In order to get to the information which learners need, they have to understand instructions and follow them, for example if they want to use a search engine, enter a MOO, open a pinboard, download software, etc. After having performed a certain task, they are immediately confronted with success or failure. In this way, surfing the Web is to some extent a continuous test of the learner's comprehension abilities.

The organisation and delivery of information through the Web is just as chaotic as in real life. Surfing the Web can probably be compared to strolling along the main street of a town in a country whose language the learners only partially speak or understand. They hear sounds, language, sometimes directed at them, sometimes not. Somebody might address them, they try to understand. Once in a while they have to ask for the way. The ultimate test whether the answer is understood is when the learners finally find the place they are looking for. Although teachers and methodologists usually recommend visits to the target language countries, they are rather wary of the unstructured, chaotic environment of the Web. They suspect that this situation and the learners' reaction to it hinders rather than furthers the learning process. Even proponents of the Web seem to have doubts on this subject:

- ... there is also no evidence that the mere locating and gathering of information improve language competence. There has to be a linguistic purpose behind the surfing, the collection of information, the exchange of information with others. (Carrier 1997:281)

- And the Web can be addictive, leading users to spend hours in aimless, unstructured browsing. (Davies 1998)

- The Internet cannot teach students to speak English. (Carrier 1997:281)

Though Felix (1999) enthusiastically claims that Web-based learning is "a window to the authentic world, she still feels the necessity to "harness" an "uncontrollably chaotic environment" for "meaningful language learning". Aimless browsing like wandering aimlessly through the streets of a foreign city, may indeed be a waste of time if we assume that all human activity needs clearly defined and structured conscious goals. It may also be a waste of time in terms of retrieving a certain piece of information. The learners, nevertheless, confronted with input on different levels, are continuously challenged to understand. They may have spent hours of unconscious language learning. Teachers and methodologists, even if technologically minded, seem to believe in the control of input. The enormous potential of the Web for unconscious learning does not seem to have impressed them.

In conclusion, while many of those involved in second language learning are convinced of the potential of the Web, two opposing views are evident in relation to its use:

i   Language learning takes place as soon as the learners enter a Web page in the language that they learn.

ii Language learning only takes place when learners have the intention to learn, when the environment as well as the task they have to perform is learner-specific, and when they are guided by an adviser/teacher/tutor.

There is ample research on second language acquisition that provides evidence that language learning for the major part is an unconscious process which can hardly be influenced by formal structured teaching. It has also been shown that under certain conditions languages can be learnt in naturalistic environment without instruction.[1] It is surely arguable that the Web represents such an environment and that in the future research in second language acquisition will increasingly be obliged to focus on the Web as a new learning context. Before going even further and considering the role of virtual reality in language learning (Schwienhorst 1998), it is essential to find out first what kind of learning takes place in these new environments and what impact the very nature of the Web has on raising the level of proficiency. There is thus an urgent need for empirical research in such areas, even though it may show that the learning processes are not very different from those with which we are already familiar.

5. Controlled environments: Web pages for learning

Web pages which have been specially designed for language learning have become so numerous that it is almost impossible to present an overview of them. Their overall purpose is to present a controlled and structured environment in which language learning is enhanced. They are usually not custom-made, i.e. specially designed for a specific learner audience, with the result that everyone looking for learning possibilities should be able to make use of them. Their overall design ranges from simple "read text and answer question" exercises, whereby the answer sheets have to be printed out and filled in by hand, to sophisticated "interactive" pages where the learners' answers are machine-checked and presented shortly after they have been entered. Their human-machine interactivity, however, is extremely limited to a few options which the learners have to chose as their answers (see, for example, http://www.wm.edu/CAS/modlang/grammnu.html). If one looks beyond the attractive design and has overcome the initial surprise that this kind of human-machine interaction really works, at least technically, one is faced with a *déjà-vu* experience in teaching methodology. Breeze hints at the essential drawback of those Web pages: "There is a legitimate concern here that thirty years of communicative methods are being thrown out of the window for the sake of virtual drill and kill" (1999:21). The learner is faced with a technologically advanced, consumer-friendly version of his textbook from the sixties, with Web pages created by designers who know more about Web design than about new methodological approaches in language teaching. As far as this kind of restricted interactivity is concerned, it should be noted that CD-ROM programs are more sophisticated and attractive, especially if they allow for direct access to authentic Web materials. Above all, they contain more data to work with and present them in a better technical quality (see Davies 1998).

In conclusion, the problems considered above are in no way peculiar to the Web, which can be seen as a mirror of a language learning and teaching market which still offers the whole range of methodological approaches from grammar-translation to task-based project-learning. Despite its apparent difference from what has gone before, it is arguable that the Web does not provide more options, but only presents them in a different form.

6. Integrating the Web: The WWW in the language classroom

Integrated into a classroom with a teacher and a specific group of learners, the Web can be either an additional tool among many others or the central one. Four different methods of integration can be distinguished:

The Web as a source of authentic texts and materials that the teacher provides for the learners

The Web is the gateway to authentic materials. It also gives teachers the opportunity to continuously update their materials and as Breeze states: "It is here, in the need for constant updating that the WWW comes into its own" (1999:23). Readers for courses in printed form will sooner or later be replaced by Web pages which provide all the texts used in the course. This is specifically useful for lesser taught languages where there is a shortage of printed materials, simply because for publishers they do not generate sufficient profit (For an example of for a Web project on Icelandic, see http://www2hu-berlin.de/bragi). Using the Web, materials can be published at reduced costs and they can be made available to all learners and enthusiasts wherever they live.

Learners are given tasks to search for information themselves

It is essential to differentiate between tasks that are teacher-initiated and those that have their basis in student interests. To take one extremely teacher-driven example, the Web page *Teaching with the Web* (http://polyglot.lss.wisc.edu/lss/lang/teach.html) presents a task where students have to access websites with information about metros in cities around the world:

> The traveller picks departure and arrival stops on the metro system in the city of their choice [...] The program calculates the approximate duration of the trip, lists all the lines taken and the stations passed. Have students report back the computed route and the monuments they stopped to visit on the way ...

Tasks like the one above are not very different from bringing newspapers into the classroom and asking the students to find articles about a topic that is given by the teacher. The Web allows for student-driven and student-initiated searches that have a much better motivational quality:

> Old school is teacher centred, who is the primary audience for student work and performance. New school incorporates projects that have intrinsic value to the students and a life beyond the confines of the classroom. (Global SchoolNet Foundation 1999).

Teachers and learners publish information for other learners or usually for a non-specified audience
In the "old" classroom writing assignments usually took the form of an exercise with little communicative intent. The teacher, often the only addressee, would be likely to grade and evaluate the form much more than the content. Publishing on the Web gives learners a chance to reach out and communicate to a much wider audience.

> In the most successful classrooms we've found, publishing on the Web is really the beginning of the story [...] for publishing on the Web opens the door of the classroom to reaction from the audience in the outside world, and provides wonderful opportunities for the Web author to dialog with that audience. (Global SchoolNet Foundation 1999)

The School of Languages at the University of Melbourne has integrated Web publication projects into their foreign language curriculum (Debski 1999). However, with more and more students publishing, the advantages of this tool are increasingly being called into question. Who reads students publications except for students, usually also learners rather than native speakers in similar projects? It seems much more effective to specify the audience right at the beginning of a project and to integrate publishing into communicative exchanges with online project partners.

The Web serves as a platform for communicative exchanges between learners and native speakers or between different groups of learners in different countries
It is here that the real strength and also the most innovative use of the Web is to be found. Whether the platform on which learners communicate is a simple pinboard system which allows the exchange of documents or a highly sophisticated environment, the major objective is to give students hands-on experience of co-operation using the language they are learning. The problem of teaching intercultural communication in the classroom is that it is almost impossible to model experiential learning in role-plays. Students who have not experienced the difficulties in co-operation with people from other cultures usually do not see the need for this type of exercise. Internet projects where they have to co-operate, to work on a common goal with learners from other cultures could eventually replace role-plays and give

the students the chance to experience authentic co-operation. Warschauer (1997) presents evidence for the fact that students actually lose interest if their published work does not have any communicative significance.

While all four options are legitimate ways of integrating the Web into the classroom, their success will ultimately depend on the teachers and their willingness to give up some of their traditional role as the omniscient possessor of knowledge. Once again, this is not a problem peculiar to Web-integration. Student-driven teaching, which was before the Web, has now become a more attractive option, allowing cross-cultural exchanges that had not previously been possible to the same extent. (See Warschauer 1997 for research on collaborative learning.)

Even the most well-intentioned, open-minded teachers, however, could be faced with a major threat to all their endeavours: the culture of the learners, i.e. their concept of what the role of the teacher should be and their expectations what the classroom should look like. It has been argued elsewhere (Vogel 1995) that teachers are sometimes under the false assumption that they share the same concept of learning and teaching with their learners. In situations where teacher and learner come from different cultures, this is most often not the case. With regard to the Web this could mean that even in situations where students enthusiastically surf the Web in their spare time, they could resist using it for formal learning, because language learning is seen as something "serious and academic", a world to which the Web with all its glossy colours and funny animations does not belong. There are indications that the reactions of some of the students in Debski's (1999) study are due to the fact that students find it difficult to reconcile the world of the Web with the world of learning:

- "...a joke going around the first couple of weeks: Oh, yea that's the hour to check your email and surf the Web for whatever you like."

- "I thought I didn't learn any new things." (Debski 1999:5)

7. Problems and further questions

The focus of this chapter so far has been on the integration of the Web into learning and teaching with particular reference to practice and practical experiments with different options. By way of conclusion, some of the theoretical implications of the issues which have arisen will be listed.

Firstly, the Web has brought new text forms and consequently new styles. Are learners being prepared for a new kind of discourse? In which way can skills be transferred to other more traditional types of discourse. Could this not lead to stylistic crossovers, e.g. learners using informal Web style in academic articles? Secondly, the design of Web pages is sometimes superior to the content. The texts are often written in poor English. To what extent will this eventually lead to a Web-pidgin? Thirdly, much has been written about the dominance of English in the Web.

According to a recent survey of the Internet Society (see Graddol 1997), about 84% of all homepages are in English. Is this an undesirable situation and how could communities of other languages exploit more fully the powers of the Web? Fourthly, if the Web presents a culture and/or subculture of its own, teachers may encounter the difficulty of having to present cultural diversity, i.e. knowledge about other cultures, through the medium of a more and more unifying Web culture. Fifthly, there are hardly any stimulating resources for beginners on the Web. In what ways could authentic sources be used for beginners? Finally, learners, continuously confronted with the information overload on the Web, could lose their empathy and finally develop some resistance to it. How do teachers achieve a balance between face-to-face communication, other media and the Web in language learning and teaching?

While pondering these questions, one should, however, not overlook the fact that in most classrooms of this world it will still take some time before the Web will make its way into language learning and teaching. In most countries of Central and Eastern Europe and in the Third World, the cassette recorder might be the only piece of technology in the classroom. For those already familiar with the use of the Web, technological advancement has not yet brought accompanying practical, methodological and theoretical solutions to the problems encountered. It seems that there are more questions than answers at present: questions that have to be answered to convince teachers that it is worth the time to think about using the Web in their teaching.

Note

[1]   See Ellis (1994) for a comprehensive overview of the field.

References

Global SchoolNet Foundation (1999). *Introduction to NetPBL: Collaborative project-based learning*: http://www.gsn.org/Web/pbl/index.htm

Breeze, R. (1999). ESP and the WWW: A framework for integration. *IATEFL ESP Special Interest Group Newsletter, 13*, 21–26.

Carrier, M. (1997). ELT online: The rise of the Internet. *ELT Journal, 51.3*, 279–309.

Collot, M. & Belmore, N. (1996). Electronic language: A new variety of English. In: S. C. Herring (Ed.), *Computer-mediated communication. Linguistic, social and cross-cultural perspectives* (pp. 13–28). Amsterdam, Philadelphia: Benjamins.

Davies, G. (1998, May). *Exploiting Internet resources off-line*. Paper presented at

the Language Teaching On-Line Conference, University of Ghent, Belgium. Available (2000) at:
http://ourworld.compuserve.com/homepages/GrahamDavies1/gdghent.htm

Debski, R. (1999). Project-based language learning and technology: An emerging alliance. Paper presented at AMEP: 50 Years of Nation Building, Melbourne 10–12 February 1999:
http://www.immi.gov.au/amep/conference/papers/debski.htm

Debski, R. & Levy, M. (Eds) (1999). *WORLDCALL: Global perspectives on computer-assisted language learning*. Lisse: Swets & Zeitlinger.

Ellis, R. (1994). *The study of Second Language Acquisition*. Oxford: Oxford University Press.

Felix, U. (1999). Web-based language learning: A window to the authentic world. In R. Debski & Levy, M. (Eds.), *WORLDCALL: Global perspectives on computer-assisted language learning* (pp 85–98). Lisse: Swets & Zeitlinger.

Ganderton, R.J. (1998). *New strategies for a new medium? Observing L2 reading on the World Wide Web*. M.A. thesis, University of Queensland:
http://www.cltr.uq.edu.au/~rogerg/thesistoc.html

Graddol, D. (1997). *The future of English?* London: The British Council.

Grassian, E. (1997). *Thinking critically about World Wide Web resources*:
http://www.library.ucla.edu/libraries/college/instruct/web/critical.htm

Herring S.C. (Ed.) (1996). *Computer-mediated communication. Linguistic, social and cross-cultural perspectives*. Amsterdam, Philadelphia: Benjamins.

Negretti, R. (1999). Web-based activities and SLA: A conversation analysis research approach. *Language Learning and Technology 3.1*, 75–87.

Osuna, M.M. & Meskill, C. (1998). Using the World Wide Web to integrate Spanish language and culture: A pilot study. *Language Learning and Technology, 1.2*, 71–92.

Rüschoff, B. & Wolff, D. (1999). *Fremdsprachenlernen in der Wissensgesellschaft. Zum Einsatz der neuen Technologien in Schule und Unterricht*. Ismaning: Hueber.

Schwienhorst, K. (1998). The "third place" – virtual reality applications for second language learning. *ReCALL, 10.1*, 118–126.

Singhal, M. (1997). The Internet and foreign language education: Benefits and challenges. *The Internet TESL Journal:*
http://www.aitech.ac.jp/~iteslj/Articles/Singhal-Internet.html

Van Lier, L. (1996). *Interaction in the language classroom*. London: Longman.

Vogel, T. (1995). Cultural determinants of foreign language learning. *Proceedings of the Eltecs Third Annual Conference, Belfast 6–9 September 1994* (pp. 81–84). Manchester: The British Council

Warschauer, M. (1997). Computer-mediated collaborative learning: Theory and practice. *Modern Language Journal, 81.3*, 470–481.

Warschauer, M. & Healey, E. (1998). Computers and language learning: An overview. *Language Teaching, 31*, 57–71.

Widdowson, H.G. (1980). *Explorations in applied linguistics*. Oxford: Oxford University Press.

10

Concordances in the classroom: The evidence of the data

Joseph Rézeau
Université de Rennes II, FR

1. Introduction

Concordancing programs have recently become more user-friendly and affordable on small computer systems; text corpora have likewise become more readily available in computer-readable format. However, such tools are still either unknown to or ignored by many language teachers. The purpose of this chapter is to give a number of concrete examples of applications of concordancers to the teaching (and learning) of English grammar at university level, and to provide the basis for a classification of concordance-based exercises.

Although the history of concordancing can be traced as far back as the 13th century, when Hugo de San Charo, assisted by 500 monks, produced a complete concordance of the Latin Bible, its use as a tool for language learning is a much more recent phenomenon, dating back to the 1980s, when computers really became increasingly powerful and affordable "word-crunchers". Still more recently, we have witnessed the concurrent emergence of computer programs permitting fast search procedures for the collocations of words in languages and availability of large corpora on CD-ROMs. Such tools and data are available to the language teacher who can use them in the classroom to promote a new type of teaching, based on authentic data, for which Johns (1991) has coined the expression data-driven learning (hereafter DDL). However, despite the growing number of papers and books on the subject of concordances, many language teachers are still either ignorant of their existence or sceptical about their potential. According to Stevens, "language teachers fall into three groups: those who have never heard of concordances, those who haven't yet taken them seriously, and those who swear by them" (1995:2). This chapter aims to examine the current situation in order to encourage those of our colleagues who either have not yet heard of concordances or who still consider them as a gadget to give them more serious consideration.

2. Why use concordances?

Widdowson (1996:68) argues that there is an inherent contradiction between the two concepts of autonomy and authenticity, as "the language which is real for native speakers is not likely to be real for learners" and concludes that "authentic language is, in principle, incompatible with autonomous language learning". Interestingly, the OED informs us that the Greek etymology of *authentic* is *authentes*: "one who does a thing himself, a principal, a master, an autocrat". Thus we see that the two concepts of authenticity and autonomy originally shared the common denominator of *auto, self.* Along with Johns (1991), this chapter will argue that DDL offers precisely the sort of setting in which the two concepts of authenticity and learner autonomy are reconciled rather than opposed. Such reconciliation is obvious in the three arguments expounded by Johns (1991:1–4) in favour of classroom concordancing: classroom concordancing uses authentic data for an authentic learning activity; the activity is learner-controlled; learning is viewed as research.

> ... the language learner is also, essentially, a research worker whose learning needs to be driven by access to linguistic data – hence the term "data-driven learning" DDL to describe the approach (Johns, 1991:2).

3. Where to find the data?

Although nowadays more and more text is made available in digital form, either on CD-ROM or through the Internet, it is still not so easy for the average language teacher to lay their hands on such material. Besides the very large corpora built up by funded research teams working at higher education institutions, there probably exists a large number of small, "home-made" *ad hoc* corpora, built over the years by language teachers using whatever material came to hand and was more or less suitable to their needs. One such example is the modest corpus which I built at the end of the 1980s, using the BBC news bulletins available as text on the French Minitel (code 3615 BBC). It was a painstaking process, which yielded a meagre 1,200 words per day. However, at the time of the Gulf War, the daily accumulation of that data allowed me to study *in vivo* the determination of the noun "war" in English (Appendix 1).

At the time of writing, the Internet looks like *the* preferred source of data to build a corpus, as it offers seemingly endless amounts of text in a wide variety of genres. It is also possible to browse through Collins' impressive *Bank of English*, which comprises 200 million words, if one pays a (fairly expensive) subscription. The most practical, albeit not the most complete, sources of data available to date are the British and American newspapers and magazines on CD-ROM. For instance, the *Times and Sunday Times* CD-ROM for the whole of 1996 comprised 44 million words. One may wonder whether it is necessary or useful to have such a

arge corpus. According to Sinclair (1991), who devotes a whole chapter to "Corpus creation", the answer is "yes", without hesitation:

> The only guidance I would give is that a corpus should be as large as possi-
> ble, and should keep on growing. [...] In order to study the behaviour of
> words in text, we need to have available quite a large number of occur-
> rences. (Sinclair 1991:18)

For ESP or EAP courses in the field of the media or French/English translation, an
ideal resource is a combination of CD-ROMS of the British Press, such as *The
Times and Sunday Times*, or *The Economist*, etc. and CD-ROMS of the French
press, such as *Le Monde* or the *AFP*. One caveat here is that, for obvious reasons of
copyright protection, a growing number of media news CD-ROM are encoded,
which makes it either extremely time-consuming or totally impossible to use them
straightforwardly for concordancing purposes.

One could also envisage scanning texts available in paper form. But, although
OCR (Optical Character Recognition) has made tremendous progress over the years,
it is still not 100% reliable, and the process of scanning, OCR processing and checking can take a lot of time and yield amounts of data which are not very large.
Resorting to this method of building a corpus should be reserved for the building of
very specialised corpora, and in cases where most or all of the data is available in
paper form only.

Another source of data for anyone interested in studying students' output in the
foreign language is precisely a corpus made out of such output. As more papers
submitted by students are in word-processor format, it is becoming more feasible to
build learner corpora. Moreover, corpus-based research into learner language is a
rapidly-expanding field, and books and papers are beginning to appear on the subject (see Gabrielli 1998; Granger 1998; Osborne 1996).

4. How to search the data?

Before the relatively recent availability of concordancing programmes, and in addition to the "heavy" programmes used on university mainframes, a number of more
modest programmes were available. A few concordance enthusiasts even created
their own concordancing macros to be run under commercial packages. Instances
of these are a paragraph concordancer in DOS and a KWIC concordancer in
WordBasic (Tribble & Jones 1997:95ff.), another KWIC concordancer called
FrameTeach (Rézeau 1988a and 1988b). The early nineties saw the apparition of
MicroConcord, a DOS concordancer which was subsequently upgraded to a Windows version re-named *WordSmith*. *MonoConc* and *MonoConc Pro* are recently
published Windows concordancers produced by Athelstan, Houston, USA.

Here are the main facilities one can expect of one of those latest concordancers:

- produces KWIC concordances (Key Word In Context)
- sorts concordances (to the left / to the right of the search-word)
- handles large data (*WordSmith* can search through corpora of tens of millions of words)
- handles tagging systems (e.g. SGML)
- produces word lists with statistical information (number of types / tokens in a given text / corpus)

For a detailed review of *WordSmith*, see Tribble (1996).

The *Collins COBUILD English collocations on CD-ROM* is not a concordancer, but a package which offers "ready-made" collocations (KWIC concordances) of 10,000 headwords making up 'the core vocabulary of English'. It can be used as a stand-alone piece of software to familiarise students with concordances, as it does not require the user to be conversant with search operations. Unfortunately, the option which was taken by its authors to ignore grammatical words (stop-words) makes it unsuitable for grammar-oriented activities such as will be discussed below.

5. What can be done with concordances?

The bulk of this contribution will consist in practical examples of using concordances with French students studying English grammar in their second year at university. Having observed that most students tend to adopt a deductive approach in their learning of grammar, to the point of thinking that language is derived from rules rather than the reverse, I have formed the hypothesis that a DDL approach to grammar, based on authentic data, might provide the learners with opportunities for a more inductive type of learning.

The remainder of this chapter will consist of a brief account of the various concordance-based exercises and activities which will be found in the appendices, and which illustrate of themselves my experimental goals.

"That" clauses (Appendix 2)

A recent trend in language teaching has consisted in denouncing the traditional division between the teaching of the lexicon and that of grammar (see Lewis 1993). Tréville & Duquette (1996:12) propose the concept of lexicon-grammar, based on the central claim that there is constant interaction between those two fields of the language. Poole (1998) uses concordances to increase his students' awareness of each verb's combinability with other lexical items. When using concordances, one grows aware that the division between grammar and the lexis is very artificial indeed, as is shown in the exercise in Appendix 2. Among the nouns which have been removed from the blocks of citations and which the students have to put back in, it is clear that a certain number function exactly like their corresponding verbs, in the context of a "that" clause. Moreover, as is often obvious in concordance-based exercises, this exercise shows that, if there are synonyms in the *lexicon* of a language, there are none in its *vocabulary*, defined as "words in use". The original idea for this exercise is to be found in Johns (1991:42).

"Much / very" followed by adjective / past participle (Appendix 3)

This exercise was prompted by Johns's view that [descriptions in traditional grammars] "are more often based on the 'armchair intuition' of the grammarian than on any close analysis of data." (1991:3). The students were asked to do their own research, based on analysis of data, in order to corroborate or invalidate the "armchair intuition" of the author of our grammar text-book. At first sight, a number of contextual clues seem to corroborate our grammarian's explanation (b): "very" is almost always used in a first-person context (the favoured person to express a "feeling", an "impression"), whereas "much" is used in second and third person contexts. On the other hand, a close analysis of the data does not enable us to find out whether "concerned" and "interested" can be considered as adjectives or verbs (in the past participle form). Moreover, one obvious feature in the data is the high frequency (100%) of a negative form in the collocations of "much + interested". Finally, Figure 1 shows the comparative frequency of use of "much / very" in the context of "concerned / interested" in a corpus consisting of 33 million words from the Times & Sunday Times CD-ROM for the year 1995 which yielded 125 collocations. It shows the low (for "interested") and very low (for "concerned") percentage of collocations of "much" as opposed to "very", a fact which our grammar-book does not mention.

"Cannot bear + N" / "To + Inf." / "V-ing" (Appendix 4)

We observe that "cannot bear" displays a relatively low frequency of occurrence in both corpora (UK and US), and that it mainly occurs in collocation with a Noun Phrase. The biggest surprise comes from the very low frequency of the -ING form (especially the NP + V-ING form). This fact contradicts the presentation of the use of "cannot bear" in several grammar-books used with my French students of English, which say that it collocates with an -ING verb form. The lesson to be learnt here is that most grammars distort reality, not only because they tend to devote more space to exceptions than to regularity in the language, but also, as in the present case, by omission. Such expressions as "can't bear" and "can't stand" are only mentioned in the context of an -ING or infinitive *verb* form, while concordance data shows that these expressions are in fact mostly followed by a Noun Phrase. Using a concordance enables us to have a statistically more reliable view of the reality of a language.

"I spent all my life / I have spent all my life" (Appendix 5)

The starting point for this concordance-based research was the following sentence which was part of a French-English translation exam: "J'ai passé toute ma vie à étudier la philosophie, mais je ne suis guère plus avancé." One colleague proposed to translate this as: "I spent all my life studying philosophy..." which I contended was either incorrect British English or possibly correct American English. The concordance data supported this view in citations 1 to 6 (Appendix 5). It is clear in these citations that the use of the simple past corresponds to a former part of the life of the speaker, a part which is clearly cut off from his/her present life (citations 3 and 6), whereas the present perfect is used for a part of the speaker's life which extends up to the time of speaking (citations 1, 2, 4 and 5). In the case of the original French sentence, it seemed that, although the speaker may be at the end of his life, he is still alive, and thus the present perfect should be used. But what about citation 7 of our data: "*I spent my whole life chasing shadows.*" in which the speaker seems to be talking about his whole life up to the present time, and nevertheless uses the simple past? A closer look at the wider context of that citation (Appendix 5:7a) enlightens us: we learn that the speaker is a transsexual whose former life (before the operation), a life "spent chasing shadows" is definitely over and cut-off from the new life. One could not wish a better proof of the power of concordances than that "authentic" citation; an example which, ironically, would look suspiciously fabricated if it were found in a grammar-book! Once more our findings are akin to those evoked by Johns:

> [the learners] will often notice things that are unknown [...] to the standard works of reference on the language. It is this element of challenge and of discovery that gives DDL its special flavour and stimulus. Johns (1991:3)

Parallel concordancing: The example of "although" (Appendix 6)
Parallel concordancing is a recent development in the field of concordancing. A Lingua funded project to explore and research this domain has produced the *MultiConcord* package, developed at Birmingham University, and tried and tested at a number of European institutions, including our own university of Rennes II. Amongst the rationale for parallel concordancers, the Birmingham team writes that:

> ... it could form the basis for a reassessment of the place of translation in general foreign language teaching - for example, in giving opportunities for "applied contrastive analysis" and in weaning students from the myth of one-to-one correspondence between first and second language. (http://sun1.bham.ac.uk/johnstf/lingua.htm)

There is no doubt that a parallel concordancer together with a corpus of parallel texts could prove an invaluable resource to supplement grammars and bilingual dictionaries.

As an example of classroom use of a parallel concordance, the exercise in Appendix 6 proposes an exploration of the many French equivalents for "although". The variety of translations found in a relatively small corpus should be enough to prove the value of this tool. The advantage of the parallel (or multilingual) corpus

over a monolingual corpus, when used to work on the grammar of the target language, is the absence of comprehension problems due to vocabulary, since the student has the translated text at his disposal. A number of similar exercises are also
available on my website (see Web references below).

6. Conclusion

I would like to start this conclusion with two caveats.

First, it should be obvious that a regular use of concordances pre-supposes easy
access to the data and concordancing programs. As I have pointed out in Section 3
and Section 4, such data and tools have now become easily available. However, a
lack of suitable hardware or software for use with the students need not be a deter-
rent for the use of concordances. Although first-hand contact with the "raw data" of
the language through manipulation of concordance packages by the students themselves may prove an enriching experience, the sheer amount of the data can be
daunting to less advanced students. On the other hand, if the teacher goes through
the first stage of sieving through the data, and selecting an unbiased but meaningful
amount to produce the sort of "worksheets" which are presented in this chapter, a
lack of resources allowing the students to handle the data directly is not a problem.

The second caveat is that one must be aware of the limitations arising from the
genre of the texts of the corpus used. A balance has to be found here between what
kind of corpus is easily available, what kind of investigation of the data is to be
made, and for what purpose. In my own experience (over more than six years),
using a corpus consisting almost exclusively of the British press has never been a
handicap. However, a more balanced corpus, such as the one produced by the BNC,
including data from spoken as well as written discourse, would prove most valuable. (See under *Websites and software* references.)

Finally, to support further the points made in this chapter, and to make even
more explicit the kind of "spirit" in which the various language activities proposed
in it have been constructed, I would like to quote from the *MicroConcord* manual:

> Whether one opts for putting up a case, or for knocking one down, any
> search using [a concordancer] is given a clearer focus if one starts out with
> a problem in mind, and some, however provisional, answer to it. You may
> decide that your answer was basically right, and that none of the exceptions
> is interesting enough to warrant a re-formulation of your answer. On the
> other hand, you may decide to tag on a bit to the answer, or to abandon the
> answer completely and to take a closer look. Whichever you decide, it will
> frequently be the case that you will want to formulate another question,
> which will start you off down a winding road to who knows where. (Murison-
> Bowie 1993:46)

It is precisely this "winding road", along which one may come across serendipity learning, which gives concordances a certain appeal. In addition, once you have started relying on the evidence of the data for checking the "rules" found in grammar-books as well as your own "intuitions" about language, concordances tend to become an indispensable tool. It is hoped that the rationale and examples given in this chapter will have convinced its readers to take a trip to the country of concordancers to observe "the company that words keep" (Firth 1957:187).

References

Firth, J. (1957). A synopsis of linguistic theory 1930–1955. In *Studies in linguistic analysis*. Oxford: Philological Society. Reprinted in Palmer, F. (Ed.) (1968). *Selected papers of J. R. Firth*. Harlow: Longman.

Gabrielli, R. (1998). Incorporating a student corpus in your teaching. *IATEFL Newsletter, 141*, February/March, 14–15.

Granger, S. (Ed.) (1998). *Learner English on computer*. London: Longman, 1998.

Johns, T. (1991). Should you be persuaded: Two samples of data-driven learning. *ELR Journal, 4*, 1–16.

Larreya, P. & Rivière, C. (1991). *Grammaire explicative de l'anglais*. Paris: Longman France.

Lewis, M. (1993). *The lexical approach*. Hove, England: Language Teaching Publications.

Murison-Bowie, S. (1993). MicroConcord Manual. Oxford: Oxford University Press.

Osborne, J. (1996). Sensibilisation lexico-grammaticale à l'aide d'un outil de concordance. *Asp, 11.14*, 417–422.

Poole, B. (1998). Corpus, concordance, combinability. *IATEFL Newsletter, 141*, 13–14.

Rézeau, J. (1998a). Que faire avec un outil professionnel en EAO des langues? *Les langues modernes, 5*, 49–60.

Rézeau, J. (1998b). De l'utilisation d'un progiciel professionnel en EAO des langues. In *Nouvelles technologies et apprentissage des langues. Le français dans le monde*, numéro spécial, 44–58.

Sinclair, J. (1991) *Corpus Concordance Collocation*. Oxford: Oxford University Press.

Stevens, V. (1995). Concordancing with language learners: Why? When? What?, *CAELL Journal, 6.2*, 2–10.

Tréville, M-C. & Duquette, L. (1996). *Enseigner le vocabulaire en classe de langue.* Paris: Hachette.

Tribble, C. (1996). Feature review: WordSmith Tools. *CALL Review, the Journal of the IATEFL Computer SIG*, September, 13–18.

Tribble, C. & Jones, G. (1997). *Concordances in the classroom.* Houston: Athelstan.

Widdowson, H.G. (1996). Authenticity and autonomy in ELT. *ELT Journal, 50.1*, 67–68.

Websites and software

British National Corpus: A very large corpus of modern British English, designed to present as wide a range of modern English as possible: http://info.ox.ac.uk/bnc

Collins COBUILD English Collocations on CD-ROM. London: HarperCollins: http://www.harpercollins.com

ICT4LT, Module 2.4: Using concordance programs in the modern foreign languages classroom: http://www.ict4lt.org/en/en_mod2-4.htm

Joseph Rézeau's Homepage: http://www.uhb.fr/campus/joseph.rezeau

Le Monde sur cédérom: Office Central de Documentation 33, rue Linné F-75005 Paris, France.

MonoConc, MonoConc Pro: Athelstan, 2476 Bolsover, Suite 464, Houston TX 77005, USA: http://www.athel.com

Tim Johns's Classroom Concordancing / DDL Bibliography: http://sun1.bham.ac.uk/johnstf/biblio.htm

Tim Johns's Data-Driven Learning Page: http://sun1.bham.ac.uk/johnstf/timconc.htm

The Times & Sunday Times on CD-ROM: News Multimedia PO Box 5060, Leighton Buzzard, LU7 7YS, UK. Tel. +44 (0) 1525 852813: http://www.newsmultimedia.co.uk

WordSmith: Oxford: Oxford University Press: http://www1.oup.co.uk//elt/catalogue/multimedia/WordSmithTools3.0

Appendix 1: A study of the determination of the noun *war* at the time of the Gulf War, or "grammar *in vivo*"

The citations below are sorted in chronological order, from 13 September 1990 to 25 January 1991.

During the pre-war period, *war* is used first with a zero degree of determination, with no article (citations 1-3, 5 & 6); but degree one of determination starts appearing, with the indefinite article (citations 4, 7, 8, 10, 12). In parallel with this determination through the use of the article, we notice another instantiation of determination, an anchor in reality, with the use of *Gulf* (6, 9, 10).

The news bulletin of the first day of the Gulf War is unfortunately missing from our corpus. However, the citation from the news bulletin of the day after (13, dated January 18, 1991) shows for the first time the use of degree 2 of determination, with the definite article *the*. During the war and post-war periods, the Gulf war is referred to according to three modes of determination:

a pre-determination: the *Gulf* war;
b post-determination: *the* war *in the Gulf*
c self-determination by the wider context : *the* war

```
1.of optimistic over-excitement". He said war remained the chief scourge of manki  900913
2.ing said their purpose was not to go to war but to see a peaceful Iraqi withdra  900918
3.sein has told his people to prepare for war. In a statement broadcast on Iraqi   900921
4.              is sending your sons to a war which has no human value or meaning, 900926
5. tions he insisted that he did not want war. "Mr Bush, ladies and gentlemen, is  900926
6.sses as big as in Vietnam if they go to war in the Gulf. In a videotaped messag  900926
7.on, also said Iraq's ability to fight a war had been over-rated. He said its ef  900926
8.lti-national deployment to be used in a war of aggression have been a complete   901120
9.  Mr Major, he said. If there is a Gulf war, Mr Major would approach it in a to  901126
10.Iraq is stepping up preparations for a war in the Gulf. Special civil defence    901213
11.on gives Congress the power to declare war, he must seek permission              901220
12.f crisis and the chances of avoiding a war. The move follows talks in Cairo yes 910103
13.ttle damage, while escalating the Gulf War into its second full day. The Israel 910118
14. in which he is expected to review the war's progress. The War Cabinet last met 910121
15.    WAR Today, MPs will debate the Gulf war but they might not vote on the issue 910121
16. t mistreatment of Allied prisoners of war. Defence Secretary Dick Cheney said  910121
17. in Kuwait have renewed fears that the war could cause serious environmental da 910123
18.    WAR The Prime Minister has said the war in the Gulf is progressing satisfac  910125
19. in London has predicted that the Gulf war will be over "in weeks". Ghazi Al-R  910125
```

Appendix 2: THAT clauses

Identify the nouns which are missing in the following blocks of citations: allegations, argument, assumption, belief, claim, conclusion, doubt, evidence, idea, impression, view

1

ngland officials recently in emphasising the current are consistent with sustain	that 6 per cent base rates
s parents, Barbara and Allan, supported the doctors' for their son, who was conde	that nothing could be done
stralia, which has inherited the traditional English best protected by the common	that freedom of speech is
ave gone out of business, reinforcing the widely held impossible to run a successful	that it is economically
s the intellectual criticism. Some analysts take the a depressing effect, and may	that the package will have

2

had drunk a similar amount. > There is also alarming smoking cigarettes, despite sustai	that more girls are
ia, all of which concluded after examining available smoking was a "health hazard	that exposure to passive
ools health education unit offers the first detailed under 11 is now widespread a	that drinking by children
n undo a great deal of the damage. "There is growing reversible and that blockages in	that heart disease is
which has taken $12m. > The greatest single piece of opening its heart to a different o	that the Academy is

3

he ground every night. But you can make a pretty good lives by dropping the bomb. When	that we actually saved
f the DNA testing on this case, so there's no logical relates to examiner bias, and th	that this, in any way,
And, of course, that might support the prosecution's with Nicole Simpson. Now, f	that he was still obsessed
n and her psychological makeup, there's pretty strong would be worst torture for her t	that life imprisonment
llings, if not both, and- They would have to make the scheming in some form or fashion	that O.J. Simpson was

4

Cocteau's work, but it succeeded in leaving a clear the charlatan he was so oft	that he was much more than
rtant for us to play it and try to correct the false away with the league.'' Wycom	that Wycombe are running
l groups indeed. All this can produce the misleading poorer results. One reason	that small classes produce
hey satisfy an aspirational need. One gets the strong we all went back to the time	that Fussell would rather
cripted "versions" of my book, but I was under the sessions had at least rammed home	that many intensive

5

e result. By refusing to change its comfortable may be run as a club with secret un	that all of public life
more sinister. This is the depressingly common writer's life somehow invalidate h	that deficiencies in the
stion: who runs the WRU? There is the erroneous affiliated bodies. If they have ul	that it is the clubs and
ual fulfilment challenges the still widely held prudish, inhibited, and often fea	that Victorian women were
ng, Allsop continued, "has given the lie to the support the most moronic abstractio	that pop songs can only

6

Federal Reserve Board for a hearing to dismiss > The Fed froze the assets	that he breached banking laws.
radition to the US of two British women who face murder plot. > Last week, a	that they were part of a cult
HE government announced an informal enquiry into health hazard yesterday, followi	that computer games are a
ustoms officers do have the power to investigate wine and beer bought on the C	that duty paid goods such as
had bombed and gassed in the past. He rejected saying that such action had t	that the raids were illegal,

7

'English civil war", comes to the astonishing as a defender of pupils' in	that this union's "credibility
tal to within a whisker of 3m. It is a foregone and, in all probability, mo	that it will surpass that level
stify the court in interfering with the judge's professional duty by the defenda	that there was a breach of
et John Heath-Stubbs wrote: "I have come to the inhabited by unpublished mino	that this country is entirely
ng through a book at a bookstore arrived at the "stupid and ignorant". If eve	that the average American is

8

rk. All rights reserved. Despite Eastman Kodak's competitors out of the complex di	that they unfairly keep
his defense witness questioned the prosecution's during the murders. <<BO>>	that they were made by a knife
ter Calls for Action, Not Words The Bosnian Serb secured by them has turne	that the UN safe haven had been
discuss his physical limitations. To refute the overpowered by O.J. Simpson.	that Ron Goldman was somehow
at the defense was promising to substantiate the golf balls out on his lawn	that O.J. Simpson was chipping

9

y.'' >bulldogs'' to reinforce the middle-class no appreciation of art or beau	that the working classes have
Tobias Smollett had in 1763, with the mistaken lung disease. Menton, where Kath	that its climate could cure
British people go out to ski on slopes with no said. "They are like maniacs	that there are any rules,'' he
s very exciting. I think you can. It's the old but to see the light outside yo	that there is light outside,
r loss of time, money, sex and spontaneity. The myth. > A study of 65 coupl	that a baby unites couples is a

10

ave exacerbated these problems and his apparent liberalisation in the long	that consumers will suffer from
heir sights on personal customers. It is beyond people to pay for the privilege o	that the banks can expect
d this view on October 9, questioning Kennedy's provoke an invasion of Cuba a	that Khrushchev was trying to
to cause suffering to an animal in the mistaken essential to good health, and an	that the eating of meat is
demonstrated that the right-wing philosophical selfish does not work as a value s	that people are innately

11

was the moment when I knew without a shadow of a fall; he was bad before we m	that I had no hand in Eamonn's
ide, I would at least give us the benefit of the prosecutors, during our trial,	that it was self-defense. The
pressly permitted.'' > There had never been any voluntarily to a mental hos	that an adult could be admitted
refusing to defend his players. There is little fast running out. > "You c	that the manager's patience is
be read in two ways. First, there can can be no the technical details of ce	that Mr George knows more about

Appendix 3: Much / very followed by adjective / past participle

Use the data provided below to check the explanations given by LARREYA (24.2.4 c. page 249) regarding:

- the choice between 'very + adjective' and '(very) much + verb'
- the use of *very* when the past participle describes a feeling of the subject

```
sometimes feverishly so. It is much concerned with the power of
able captain, he is probably as much concerned with the creation
lds of London Institute. He was much concerned with technical edu
rsity Women's Club >Sir, We are much concerned by the plight of w
g picture of a just society, as much concerned with the needy as
e also said the public were not much interested in Europe a point
nario seem real, but he was not much interested in aliens or spac
o by his mother, but he was not much interested in music until a
rally speaking, have never been much interested in welfare. What
he Spanish border guard was not much interested in my passport bu
the engineering faculty was not much interested in his ideas, kee
well be that Banderas is not so much interested in love, love, lo
reparations, said he was not so much interested in defending his

on, a shareholder, said: "I am very concerned that the company i
h people being ripped off. I am very concerned the government has
: "Milosevic no longer appears very concerned whether a comprehe
    "The PoWs and internees are very concerned that the statement
grandson says. "My father was very concerned I was doing these
e.'' > Mrs Beckett said: "I am very concerned about this. We wil
Health Secretary, said she was very concerned and would raise qu
car Mold. "He said that he was very concerned for the rest of th
The Foreign Office said it was very concerned by reports of poli
Fund for Children said: "I am very concerned we may not qualify
he OPW, said his department was very concerned about the state of
hose approached are "fairly or very concerned" about being made

om Mr S.G. Marriott >Sir, I was very interested to read the artic
coached. Outside the group, I'm very interested in how Wales will
out their existence. I am still very interested in assisting with
. "I am not a Hitler fan. I am very interested in German history
' she explained. Her mother was very interested to hear that Dian
larly appeals to me. I would be very interested in including them
ammes, though, and everyone got very interested in the OJ Simpson
the brightest in the school and very interested in politics. He 1
to the very end. She was always very interested in politics and t
orth (Conservative) >Sir, I was very interested in Anthony Howard
```

*Figure 1: Compared frequencies of the use of **much** and **very***

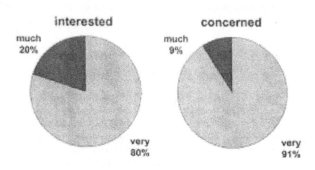

Appendix 4: Cannot bear + N / TO+INF / V-ING

UK Corpus = *Times & Sunday Times* Jan–Jun 1995 : 22,391,166 words
US Corpus = *Broadcast News*: TV and radio transcripts: 14,644,686 words

Note: the lists of citations below are not selections but the complete data returned by the following search criteria: *cannot / can't* + Noun Phrase (noun or pronoun) and a Verb Phrase either in the infinitive or the gerund

UK CORPUS

1. er, the vacuum cleaner, it all does it. I can't bear **being** kept waiting, without somebody letti
2. styling but it must have a new engine. I cannot bear **cars** which do not start every morning. Fa
3. your starting point is the fact that you cannot bear **each other**, it is not helpful to be sent
4. s having to read out work in progress and cannot bear even **his wife to look** over his shoulder.)
5. confessed, she forgave him, but now she can't bear **him touching her**. It's a terrible blow to t
6. they wail. What these narcissistic souls can't bear **is that** there exists an age, a time, anyth
7. skirts round the telephone table next. I can't bear **it**. >In fairness, this incipient mid-life
8. tyle. I switch off the air conditioner; I cannot bear **it**. I walk to a bathroom and almost canno
9. oss the Atlantic? If you could, I would I cannot bear **long flights**. Imagine going somewhere suc
10. names. Several names argue that Lloyd's cannot bear **the weight** of the losses of the past few y
11. bad word, perhaps two or more; those who cannot bear **the thought** of anyone besmirching Credit
12. disparate origin. But frankly those who can't bear **the idea** of the new, the different, the for
13. ing, but I paid a high price. Even now I can't bear **the thought** of going away. I look upon Lon
14. arried is that right?'' "Spot on, boss. Can't bear **the sight** of them at least she can't. She
15. er nostalgic for that Sloaney froideur. I can't bear **the way** everybody is suddenly so desperate
16. be getting to the point that they simply cannot bear **the thought** of him in the Oval Office for
17. not watch; will not contemplate victory; cannot bear **the tension** in the build-up to a race. >
18. re room in the house: but today's buyers cannot bear the thought of a short walk, possibly in t
19. much more than either of them realises, cannot bear **the thought** of telling her the truth afte
20. care to win it anyway expression. If you can't bear **three hours** of showbiz flatulence in the e
21. travelling journalist, an old woman who can't bear **to tell** her son about the most important e
22. oint of paralysis by working girls, they cannot bear **to admit** that they are not earning. But wh
23. e fits of terror. Even as a grown man he cannot bear **to be left** alone at night, or travel alon
24. Meanwhile, those who live in America and cannot bear **to be out of touch** with sports scores and
25. Family Therapy (BBC2, Sunday, 9.05pm). I can't bear **to reveal** what this is a real-life insight
26. : the Hidden God, and so forth but no, I can't bear **to speculate**. Either way, none of this qui

27. um of 20 minutes,'' Mrs Levine says. "I can't bear **to see boredom** in people's faces." > The

28.huffled off, no more ado. > Lloyd Webber cannot bear **to fail.** When he wasn't happy with the pro

29. an iron roof by Charles Eiffel which he cannot bear **to sell,** or the dealer in the Vendee who h

30.a couple which let in water but which we cannot bear **to throw away.** Samuele Mazza has more than

31.rliamentary magazine. "If daddy or mummy cannot bear **to be a million miles away** from Charlie o

32. y who has not read that story? Why no, I can't bear **to go on.** Apart from Tim Hatley's brilliant

33.d. This is a great comfort to owners who cannot bear **to see their pets suffer** but will not hav

34.th cold limbs, no thirst with the fever, cannot bear **to be touched,** throbbing hammering head b

35.activities. Life must go on. > Humankind cannot bear very much high **history.** The King is paying

36.y were for the first spin. But humankind cannot bear very much of that absurd **mathematics.** > M

37. who is in two minds about a suitor, "I cannot bear **you should be unhappy about him.** Think of

38.d by his own actions to endure **a life** he cannot bear. His reading is well paced and moving, an

US CORPUS:

1. litician and as a public servant, but she can't bear not **to be** with you. <<IT>><<BO>>E. JEAN CAR

2.illenium'<<NM>>: It's just incredible. We cannot bear **that burden.** Nobody can bear that burden..

3.r considers her life precious, that Smith cannot bear **to be alone,** and a life sentence will leav

4. <<BO>>CHARLIE ROSE<<NM>>: I think that she can't bear **to be away from you,** and I also think- <<IT

*Figure 2: Breakdown of collocations of **cannot bear***

	clause	Noun	Noun + V-ING	V-ING	N + infinitive	Infinitive
The Times	2	19	1	1	1	14
percentage	5%	50%	3%	3%	3%	37%
US News		1			3	
percentage	0%	25%	0%	0%	75%	0%
TOTAL	2	20	1	1	4	14
percentage	5%	48%	2%	2%	10%	33%

Appendix 5: I spent all my life / I have spent all my life?

Since the life of the speaker is not achieved, the Present Perfect should be used, not the Simple Past …

1 *I've spent a working life affecting not to mind about my degreelessness - …*
2 *I've spent a life trying to eliminate suffering from the world*, yet I find myself writing about it …
3 *"I spent most of my life in submarines, but now* I work for a cat," he said, proudly …
4 A key element in that world is *the creative environment that I have spent most of my life fighting for*, hoping for a society which values not Heritage, but culture.
5 *I have spent most of my life fighting for the liberation of the people.*
6 *I spent my entire **early** life enduring jokes about my surname* 'Callow youth,…
7 *I spent my whole life chasing shadows.*
7a SOURCE: The Times DATE: 01 April 1995 PAGE: TM/37
Just Like A Woman; Gender Crossing Chrissy Iley
Rachel sits with us, fortyish, Marks & Spencerish. She is a post-operative transsexual. […]
"I was a difficult child because I was so unhappy. In adolescence, I was supposed to be chasing girls, but found that wasn't right. I had no homosexual leanings either. But I knew from when I was a small child that I just didn't listen to myself. *I spent my whole life chasing shadows.* "I really am in love with life, whereas before, even if I went out and had a good time, I would always come back to a void."

Appendix 6: Parallel concordancing, the example of *although*

.The *Oxford-Hachette bilingual dictionary* classifies the uses of *although* in two categories and gives the following translations:

i *in spite of the fact that*   **bien que / quoique**
ii *but, however*   **bien que / mais**

The *Harrap's Shorter* adds two translations:

i *although I am a father*   **tout** père **que** je suis;
ii *although not beautiful, she was attractive*   **sans** être belle elle plaisait
   (contexte négatif)

In the following citations, try to classify each use of *although* in one of the two categories proposed by *Oxford-Hachette*. Make a list of all the other translations not proposed by the two dictionaries (including the cases where *although* is not translated by lexical but by syntactic means.

1. European institutions have only been interested in transport since 1974, **although** it was within the competence of the Treaty of Rome.

1. Les institutions européennes ne se sont intéressées au transport qu'à partir de 1974, **bien que** ce fût dans la logique du traité de Rome.

2. Ptolemy recognised this flaw, but nevertheless his model was generally, **although** not universally accepted.

2. Ptolémée était conscient de ce défaut mais son système n'en fut pas moins généralement, **si ce n'est** universellement, adopté.

3. He had a fine accordion which he hardly ever played, **although** the story went that he played wonderfully.

3. Il possédait un bel accordéon, mais n'en jouait presque jamais, le bruit courait **néanmoins** que c'était un virtuose;

4. "Hist," said Drogo two or three times timidly to attract the attention of the apparitions, **although** he knew quite well in his heart that it would be useless.

4. "Psst... psst...", fit timidement Drogo deux ou trois fois, pour attirer l'attention des fantômes, **tout en** sachant bien néanmoins, au fond de son cœur, que ce serait inutile.

5. Night had fallen, **although** up to now no one had thought seriously of it.

5. La nuit était tombée, la nuit à laquelle jusqu'alors personne n'avait sérieusement pensé.

6. But Drogo had not seen it, **although** he had tried hard enough.

6. Mais Drogo ne l'avait pas vue, **malgré** tous ses efforts.

7. Many of them said that it really was a road **although** they could not say what it was for...

7. Beaucoup disaient que c'était vraiment une route, sans réussir **pourtant** à en expliquer le but;

8. But for the men it has been a long road **although** they do not quite understand how it passed so quickly.

8. Mais pour les hommes, elles ont été un long chemin à parcourir, **encore que** l'on ne comprenne pas comment elles ont passé si vite.

9. And **although** the mysterious tumult of the passing hours grows with each day, Drogo perseveres in his illusion that the really important things of life are still before him.

9. Et l'angoisse obscure des heures qui passent **a beau** se faire chaque jour plus grande, Drogo s'obstine dans l'illusion que ce qui est important n'est pas encore commencé.

10. Simeoni had imagined that Drogo could not move any more, had paid no more attention to him, had taken decisions on his own, **although** of course he would tell him when everything had been done.

10. Simeoni s'était figuré que Drogo ne pouvait pas bouger, il ne s'était plus soucié de lui, il avait pris des décisions tout seul, **quitte à** l'informer ensuite quand tout aurait été exécuté ...

11. "Listen," replied Drogo, **although** he saw how absurd it was to fight on;

11. "Ecoute", répondit Drogo, **mais** il comprenait combien il était absurde de lutter;

12. Europe was favoured, **although** represented rather awkwardly, with improbable disproportions.

12. L'Europe se trouvait privilégiée, **quoique** représentée assez gauchement avec des disproportions invraisemblables.

13. **Although** we do not know how he came, we do know that eight years later one of his countrymen,

13. **Si** l'on ignore comment ce dernier était venu, on sait en revanche que, huit ans plus tard, un de ses compatriotes,

14. ... the land routes preserved an active role, **although** less substantial than Jan Van Houtte believes.

14. ... les voies terrestres ont conservé un rôle actif, moins prépondérant **cependant** que ne le pense Jan Van Houtte.

15. **Although** different, the types of harbours showed aspects which corresponded more or less faithfully to a model that has become commonplace...

15. **Même** différents, les types portuaires présentaient des aspects correspondant plus ou moins fidèlement à un modèle devenu banal...

16. **although** that did not yet create a Europe, nevertheless through negative paths and a convergence of contradictory interests it prepared one...from afar.

16. cela ne faisait pas encore une Europe, **convenons-en**, mais, par des voies négatives et une convergence d'intérêts contradictoires, la préparait...de loin.

17. **Although** they are not all poets, seamen themselves even now bear witness about the sea which veracity renders even stronger.

17. **Pour ne pas** être tous poètes, les hommes de la mer eux-mêmes apportent, encore maintenant, des témoignages sur la mer que la véracité rend plus forts.

11

ReLaTe: A case study in videoconferencing for language teaching

John Buckett & Gary Stringer
University of Exeter, UK

1. Introduction

This paper describes work which was being carried out as part of an investigation into the uses of multimedia conferencing for language teaching. The project, called ReLaTe (standing for REmote LAnguage TEaching), was undertaken jointly by the University of Exeter and University College, London. It was originally funded by BT as part of the UK's BT/JISC SuperJANET initiative and subsequently under the JISC's JTAP programme. The project aims to demonstrate the feasibility of using multimedia conferencing to share language teaching resources. For mainstream language teaching, it offers the potential of access to teaching and lecturing staff which would not otherwise be possible; for example, someone giving a French course in a provincial university for an audience of architects could offer a specialist seminar by a French architect based in London. For minority languages or specialisms, the advantage may be even greater; it may simply not be possible to get a viable audience for Polish at any one place, but by making a course available by videoconferencing in many locations, this problem can be overcome. In the current phase of ReLaTe, this year's trials are already taking advantage of these benefits: four prospective French students at Exeter who were unable to follow a local course because of timetable difficulties are receiving half of their tuition from a tutor based at UCL. More generally, in any situation involving distance learning, videoconferencing potentially has a very important role to play. For a typical "lonely learner", where a student is perhaps working from books, CD-ROMs and Internet materials in the evenings after finishing a full-time job, a very high level of motivation and drive is needed to keep going, with only occasional stimulus from meetings with a tutor or fellow students. In this situation, the interaction coming from regular small group videoconferencing, as demonstrated in ReLaTe, can be a very powerful motivational driver. Other forms of driver can also be seen coming into play: the "embarassment factor" of having to face your tutor when you haven't prepared a piece of work in time for a tutorial, seems to work as strongly with a videoconferenced tutorial as a face-to-face one.

This description of the ReLaTe project is presented here as a case study of the use of videoconferencing in languages and as an example of the research and development which is necessary before this type of work can become a routine part of language teaching. Before going on to look at the ReLaTe project in more detail, however, it may be helpful to put the project into a broader context.

2. Varieties of videoconferencing

There are a number of different underlying technologies in use for videoconferencing and without going into technicalities, it may be helpful to review the main categories and their potential for language teaching. One possible division is a breakdown into four categories:

2.1 Traditional studio-based videoconferencing
This relies on fixed, TV-style, studio equipped with cameras, microphones and lighting. This type of system is often linked by high-speed, dedicated communication lines and the quality can be extremely good, with VCR-grade (or better) output to large monitors. This type of videoconferencing has a long tradition in higher education (UK examples include Livenet in the University of London and Welshnet linking University of Wales higher education institutions), but the capital cost of the studios, plus the communications network rental can be relatively high. The Scottish MANs videoconferencing network (SMVCN), implemented on ATM networking technology, also offers this type of studio-based conferencing; see Mercer and Provan (1999).

2.2 ISDN-based videoconferencing
ISDN (Integrated Services Digital Network) is a digital telecommunications technology that has been in existence for many years, but only relatively recently has it become widely used for videoconferencing. The basic level of ISDN connection (ISDN 2) gives a medium-speed connection, offering a total of 128 kbps bandwidth. This type of videoconferencing can be used both for studio-based systems and on desktop-videoconferencing; on the latter, the hardware normally exists as a plug-in board which slots into a standard high-end PC. On current PC-based systems, the constrictions of both the bandwidth and processor power mean that output is some way below broadcast quality, both is terms of resolution and the number of frames per second. Considerable improvements in quality are possible by using multiple ISDN lines for a single videoconference; ISDN 6, for example, gives 384 kbps bandwidth. This ability to provide improved bandwidth is most often used when ISDN technology is used in studio-based videoconferencing, but the improvement is at the expense of call costs being increased by a factor of three.

In most simple ISDN conferences, connections are simply made end-to-end between two participants. If more than two participants wish to take part in an ISDN videoconference, then a central switch is needed to route all the calls. This type of switch is available commercially, but rental charges for the switch may be high, in addition to the increasing size of telephone charges as the number of participants increases. However, in some countries, such switches may be available at low cost for the educational community; for example, in the UK a service of this type is provided by UKERNA (JANET 2000).

2.3 Internet-based videoconferencing
The potential of using the Internet to provide videoconferencing has been apparent for some years and a number of attempts have been made to demonstrate its feasibility. Probably the most widely-known products are *NetMeeting* and *CUSeeMe*, which have both been available for PC hardware for several years. The disadvantage of many Internet connections is that they cannot reliably deliver sufficient bandwidth to provide usable conferencing. The development of the H.323 standard (Down 1999), which is used both by PC-based and larger studio systems, has given an added boost to Internet videoconferencing, and both hardware and software MCUs (Multipoint Control Units) are now available which can make multiway conferences possible.

Figure 1: A typical ReLaTe session

2.4 MBone videoconferencing
The MBone (short for Multicast Backbone) is an overlay network that enables applications such as multimedia to be sent efficiently over the Internet. The technical details are available elsewhere (Buckett et al., 1995), but it provides an efficient way for multiway videoconferences to be handled over the Internet. The ReLaTe project is just one example of an application making use of the MBone. The basic software tools (for video, audio and shared workspace) have traditionally been written by research groups for the Unix operating system. This software has been made

freely available, and help and support in its use has in the past been provided in most European countries via the MICE (Multimedia Integrated Conferencing for European Researchers) National Support Centres; in the UK the most up-to-date archive is hosted by UCL (UCL 2000). In addition to the original Unix platform, this software is now also becoming available on PC hardware, under both Windows 95 and Linux operating systems.

3. Conferencing in ReLaTe

Figure 1 shows a participant in a language tutorial being run as part of the ReLaTe project. From an early stage, ReLaTe has used headsets, with combined microphone and earphones. The use of headsets means that there are no problems of feedback from speakers to microphones, so that it becomes possible to set up the software in such a way that everybody in a conference is allowed to speak simultaneously. Experience in ReLaTe trials has shown that this is extremely important; in a language teaching context, paraverbal responses (such as "uh-huh" or "umm") are important in providing feedback as to whether students have understood.

Figure 2 shows a detailed screenshot taken during a typical ReLaTe session. The project has developed a simple-to-use interface that calls up the separate software tools (for video, audio and shared workspace). Up to four participants can take part in a conference and their video is shown in small video windows at the bottom left of the screen. A participant can click on any of the small video windows (including the one showing himself or herself) and can display it as the larger picture at top left; often, as in this example, the tutor's image is displayed in the larger window. A "powermeter" at the side of each video window indicates the audio level and thus gives a visual confirmation of who is currently speaking. The audio controls for setting levels are at the centre bottom of the screen, but most of the screen area is taken up by the shared workspace. Text and graphics, which are visible to all participants, can be imported onto this shared workspace at the start of the session. Tutor and students are all able to add text or draw in the workspace throughout the session; colour allows easy identification of the annotations by their contributor.

4. Language teaching trials

Several series of language teaching trials were carried out as part of ReLaTe, using the UK's JANET MBone service over a network which at the time provided a basic 10 Mb/s bandwidth between Exeter and UCL. The conclusion reached was that it was certainly feasible to teach foreign languages over a network such as the JANET Mbone, but that much work remains to be done before it can become a routine part of language teaching. The fundamental problem remains the variable nature of audio and video quality. Over the MBone, videoconferencing transmissions have no priority over computer network traffic; consequently, if any part of the network becomes busy, then the quality of the videoconference suffers. In practice, loss of quality on the video transmissions could be tolerated to a considerable extent. If a video picture temporarily flickered or broke up, then this caused relatively few problems. Conversely, interference with the audio was immediately felt by the participants; unsurprisingly, a language learning context is particularly susceptible to parts of the conversation being delayed or lost due to networking congestion. Nevertheless, the lessons nearly always continued even when audio was perceived as being poor and tutors and students developed strategies to reduce the problems.

Figure 2: The ReLaTe interface

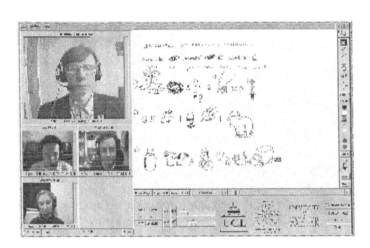

Because of the perceived importance of audio for language learning, considerable effort in ReLaTe has gone into the development of tools with audio/video synchronisation, which was not present in early MBone software tools. However, due to network limits and processor speed restrictions, most classes have been run with slow video frame rates which precluded effective synchronisation, apparently without interfering fundamentally with the teaching process. Some research elsewhere has suggested that for some tasks, video may be an irrelevance. However, experience from ReLaTe has shown that both tutors and students do value it; crucially, it provides a way of gauging reactions (e.g. frowning, smiling, puzzlement), of clarifying meaning (e.g. by mime) and as a way of learning some of the non-verbal gestures relevant to the language being taught. Interestingly, when some students met their tutor face-to-face for the first time at the end of earlier trials, they commented that they felt they already knew him as a personality.

One area of technical development from which ReLaTe has benefited has been the development of an improved audio tool, known as RAT (Robust Audio Tool). Developed by researchers at UCL, it builds redundancy into the transmission of the audio stream, so that a second, low bit-rate version of all the audio is encoded and sent, separately from the primary audio. If parts of the primary audio are delayed or lost, then the missing sections are replaced by the secondary audio stream; this technique provides greatly improved resilience for the audio in conditions of network congestion and has proved to be a major advance.

One surprising result of the trials was the very high importance that the shared workspace came to have in nearly all language classes. The ability to scan in texts and then use these as a basis for all sorts of language work (conversation, grammar classes, writing, spelling, pronunciation, vocabulary, etc.) was a major bonus. Students commented that in traditional classes, the whiteboard is the lecturer's province. The ability for everybody to have equal access to the shared workspace and write comments, underline passages etc., was found to be very liberating in many kinds of classes.

5. Future potential

In the specific field of language teaching, it already seems clear that ReLaTe has shown the vast potential of MBone videoconferencing. Since the MBone is an international network, the enticing prospect for language teaching, of course, is the prospect of running videoconferencing with other European countries. This offers not only access to lecturers and tutors based abroad, but also student-to-student contact. The prospect of a group of French or German students spending half a session practising English with their English student counterparts and then reversing roles for the second half of the lesson is a beguiling one, not least for language departments hard pressed to find sufficient teachers to teach the requisite number of contact hours. ReLaTe has recently embarked on some limited trials with the University of Würzburg, taking advantage of the increased bandwidth resulting from the setting-up of the EU-funded TEN-155 network (DANTE, 2000) which links European research networks, and the results could have a major impact on the demand for European bandwidth. However, although successful trials have taken place, multicast is currently still not a fully-supported service on DANTE and until this is the case, progress with international MBone videoconferencing will consequently be limited.

Taking our experience with ReLaTe as an indicator for the prospects for videoconferencing more generally, it looks so far as though the growth of MBone applications has the potential to take off very rapidly in the fairly short-term future. It may be that the lack of MBone support will prevent this happening as quickly as the technical developments allow. Nevertheless, computer network bandwidth is becoming cheaper and in particular, the growth of affordable high speed network connections to the home, using technologies such as ADSL (Asymmetric Digital Subscriber Line) also mean that opportunities for all types of Internet videoconferencing will expand greatly in the immediate future. With technological developments such as these in train, videoconferencing should become an increasingly important tool for language learners in the years to come.

References

DANTE (2000). *TEN-155: The current European research networking*:
http://www.dante.org.uk/ten-155.html

JANET (2000). *JANET videoconferencing services*:
http://www.jvcs.video.ja.net

UCL (2000). *MBone conferencing applications*:
http://www-mice.cs.ucl.ac.uk/multimedia/software

Buckett, J., Campbell, I.L.C., Watson, T.J., Sasse, M.A., Hardman, V. & Watson, A. (1995). *ReLaTe: Remote language teaching over SuperJANET: Proceedings of Networkshop 23, Leicester, 28–30 March 1995* (pp. 209–213). Didcot: UKERNA.

Down, P. (1999). *Videoconferencing standards*:
http://www.video.ja.net/misc/stan.html

Mercer, D. & Provan, F. (1999). *A review of the Scottish MANs' videoconferencing network*:
http://www.ja.net/development/video/manvc/SMVCN_Review.pdf

Wood, G. (1998). *The multimedia conferencing applications archive*:
http://k2.avc.ucl.ac.uk/mice

Name Index

Subject Index